ADVANCING AERONAUTICAL SAFETY

A REVIEW OF NASA'S AVIATION SAFETY-RELATED RESEARCH PROGRAMS

Committee for the Review of NASA's Aviation Safety-Related Programs

Aeronautics and Space Engineering Board

Transportation Research Board

Division on Engineering and Physical Sciences

NATIONAL RESEARCH COUNCIL
OF THE NATIONAL ACADEMIES

THE NATIONAL ACADEMIES PRESS
Washington, D.C.
www.nap.edu

THE NATIONAL ACADEMIES PRESS **500 Fifth Street, N.W.** **Washington, DC 20001**

NOTICE: The project that is the subject of this report was approved by the Governing Board of the National Research Council, whose members are drawn from the councils of the National Academy of Sciences, the National Academy of Engineering, and the Institute of Medicine. The members of the committee responsible for the report were chosen for their special competences and with regard for appropriate balance.

This study is based on work supported by NNH05CC15C between the National Academy of Sciences and the National Aeronautics and Space Administration. Any opinions, findings, conclusions, or recommendations expressed in this publication are those of the authors and do not necessarily reflect the views of the agency that provided support for the project.

International Standard Book Number-13: 978-0-309-15793-3
International Standard Book Number-10: 0-309-15793-5

Cover: Cover design by Tim Warchocki.

Copies of this report are available free of charge from:

Aeronautics and Space Engineering Board
National Research Council
500 Fifth Street, N.W.
Washington, DC 20001

Additional copies of this report are available from the National Academies Press, 500 Fifth Street, N.W., Lockbox 285, Washington, DC 20055; (800) 624-6242 or (202) 334-3313 (in the Washington metropolitan area); Internet, http://www.nap.edu.

Printed in the United States of America

THE NATIONAL ACADEMIES

Advisers to the Nation on Science, Engineering, and Medicine

The **National Academy of Sciences** is a private, nonprofit, self-perpetuating society of distinguished scholars engaged in scientific and engineering research, dedicated to the furtherance of science and technology and to their use for the general welfare. Upon the authority of the charter granted to it by the Congress in 1863, the Academy has a mandate that requires it to advise the federal government on scientific and technical matters. Dr. Ralph J. Cicerone is president of the National Academy of Sciences.

The **National Academy of Engineering** was established in 1964, under the charter of the National Academy of Sciences, as a parallel organization of outstanding engineers. It is autonomous in its administration and in the selection of its members, sharing with the National Academy of Sciences the responsibility for advising the federal government. The National Academy of Engineering also sponsors engineering programs aimed at meeting national needs, encourages education and research, and recognizes the superior achievements of engineers. Dr. Charles M. Vest is president of the National Academy of Engineering.

The **Institute of Medicine** was established in 1970 by the National Academy of Sciences to secure the services of eminent members of appropriate professions in the examination of policy matters pertaining to the health of the public. The Institute acts under the responsibility given to the National Academy of Sciences by its congressional charter to be an adviser to the federal government and, upon its own initiative, to identify issues of medical care, research, and education. Dr. Harvey V. Fineberg is president of the Institute of Medicine.

The **National Research Council** was organized by the National Academy of Sciences in 1916 to associate the broad community of science and technology with the Academy's purposes of furthering knowledge and advising the federal government. Functioning in accordance with general policies determined by the Academy, the Council has become the principal operating agency of both the National Academy of Sciences and the National Academy of Engineering in providing services to the government, the public, and the scientific and engineering communities. The Council is administered jointly by both Academies and the Institute of Medicine. Dr. Ralph J. Cicerone and Dr. Charles M. Vest are chair and vice chair, respectively, of the National Research Council.

www.national-academies.org

OTHER REPORTS OF THE AERONAUTICS AND SPACE ENGINEERING BOARD

Capabilities for the Future: An Assessment of NASA Laboratories for Basic Research (Laboratory Assessments Board with the Aeronautics and Space Engineering Board [ASEB], 2010)
Defending Planet Earth: Near-Earth Object Surveys and Hazard Mitigation Strategies (Space Studies Board [SSB] with ASEB, 2010)

America's Future in Space: Aligning the Civil Space Program with National Needs (SSB with ASEB, 2009)
Approaches to Future Space Cooperation and Competition in a Globalizing World: Summary of a Workshop (SSB with ASEB, 2009)
An Assessment of NASA's National Aviation Operations Monitoring Service (ASEB, 2009)
Final Report of the Committee for the Review of Proposals to the 2009 Engineering and Physical Science Research and Commercialization Program of the Ohio Third Frontier Program (ASEB, 2009)
Fostering Visions for the Future: A Review of the NASA Institute for Advanced Concepts (ASEB, 2009)
Near-Earth Object Surveys and Hazard Mitigation Strategies: Interim Report (SSB with ASEB, 2009)
Radioisotope Power Systems: An Imperative for Maintaining U.S. Leadership in Space Exploration (SSB with ASEB, 2009)

Assessing the Research and Development Plan for the Next Generation Air Transportation System: Summary of a Workshop (ASEB, 2008)
A Constrained Space Exploration Technology Program: A Review of NASA's Exploration Technology Development Program (ASEB, 2008)
Final Report of the Committee for the Review of Proposals to the 2008 Engineering Research and Commercialization Program of the Ohio Third Frontier Program (ASEB, 2008)
Final Report of the Committee to Review Proposals to the 2008 Ohio Research Scholars Program of the State of Ohio (ASEB, 2008)
Launching Science: Science Opportunities Provided by NASA's Constellation System (SSB with ASEB, 2008)
Managing Space Radiation Risk in the New Era of Space Exploration (ASEB, 2008)
NASA Aeronautics Research: An Assessment (ASEB, 2008)
Review of NASA's Exploration Technology Development Program: An Interim Report (ASEB, 2008)
Science Opportunities Enabled by NASA's Constellation System: Interim Report (SSB with ASEB, 2008)
United States Civil Space Policy: Summary of a Workshop (SSB with ASEB, 2008)
Wake Turbulence: An Obstacle to Increased Air Traffic Capacity (ASEB, 2008)

Building a Better NASA Workforce: Meeting the Workforce Needs for the National Vision for Space Exploration (SSB with ASEB, 2007)

Decadal Survey of Civil Aeronautics: Foundation for the Future (ASEB, 2006)
Review of the Space Communications Program of NASA's Space Operations Mission Directorate (ASEB, 2006)

Government/Industry/Academic Relationships for Technology Development: A Workshop Report (ASEB, 2005)
Review of Physical Science Proposals to the State of Ohio's Wright Centers of Innovation Award Program: Final Letter Report (ASEB, 2005)
Technology Pathways: Assessing the Integrated Plan for a Next Generation Air Transportation System (ASEB, 2005)

Letter Report from the Panel for the Review of Proposals for NASA's Intelligent Propulsion Systems Foundation Technology (PSFT) Program (ASEB, 2004)
Stepping-Stones to the Future of Space Exploration: A Workshop Report (ASEB, 2004)
Systems Integration for Project Constellation (ASEB, 2004)

Limited copies of ASEB reports are available free of charge from

Aeronautics and Space Engineering Board
National Research Council
The Keck Center of the National Academies
500 Fifth Street, N.W., Washington, DC 20001
(202) 334-2858/aseb@nas.edu
www.nationalacademies.org/aseb

COMMITTEE FOR THE REVIEW OF NASA'S AVIATION SAFETY-RELATED PROGRAMS

H. NORMAN ABRAMSON, Southwest Research Institute (retired), *Chair*
ROBERT T. FRANCIS, Zuckert, Scoutt, and Rosenberger, L.L.P., *Vice Chair*
ELLA M. ATKINS, University of Michigan
DEBORAH A. BOEHM-DAVIS, George Mason University
JAMES BURIN, Flight Safety Foundation
COLIN G. DRURY, State University of New York at Buffalo
R. JOHN HANSMAN, JR., Massachusetts Institute of Technology
PIERRE T. KABAMBA, University of Michigan
JAMES R. KRODEL, Pratt and Whitney
RAYMOND R. LaFREY, Massachusetts Institute of Technology Lincoln Laboratory (retired)
EDMOND L. SOLIDAY, United Airlines (retired)
CORNELIA TOWNSEND, Boeing Commercial Airplanes
ERIC J. TUEGEL, Air Force Research Laboratory
RAYMOND VALEIKA, Independent Consultant
WILLIAM WHITTON, Gulfstream Aerospace Corporation

Staff

PAUL JACKSON, Program Officer, Aeronautics and Space Engineering Board, *Study Director* (from August 2009)
THOMAS R. MENZIES, JR., Senior Program Officer, Transportation Research Board
BRIAN DEWHURST, Program Officer, Aeronautics and Space Engineering Board (through August 2009)
ANDREA REBHOLZ, Program Associate, Aeronautics and Space Engineering Board

MICHAEL H. MOLONEY, Director, Aeronautics and Space Engineering Board

Preface

Section 305 of the NASA Authorization Act of 2008 directed the National Aeronautics and Space Administration (NASA) to contract with the National Research Council (NRC) for an "independent review of NASA's aviation safety-related research programs." The act calls for a review of whether the programs have well-defined, prioritized, and appropriate research objectives; resources have been allocated appropriately among these objectives; the programs are well coordinated with the Federal Aviation Administration (FAA) and other relevant agencies; and suitable mechanisms are in place for transitioning the research results in a timely manner. The specific language of the study charge is contained in Appendix A of this report.

To conduct the review, the NRC's Aeronautics and Space Engineering Board (ASEB), in conjunction with the Transportation Research Board (TRB) of the National Academies, assembled a committee of 15 experts with a wide range of expertise and perspectives. H. Norman Abramson, executive vice president (retired) of the Southwest Research Institute, chaired the committee, which included experts in aviation safety, aircraft manufacturing, airline operations, aircraft aging and condition, flight control, aviation software safety, traffic operations, weather, human factors, and the Next Generation Air Transportation System (NextGen). Biographical sketches of the committee members can be found in Appendix B of this report.

Over the course of 9 months, the committee met four times. During its first meeting on June 22-23, 2009, the committee received overview briefings on NASA's safety-related research from Amy Pritchett, then-director of NASA's Aeronautics Research Mission Directorate's (ARMD) Aviation Safety Program. To better understand its charge, the committee also talked with Richard Obermann, a member of the staff of the House Committee on Science and Technology. In addition, the FAA's Barry Scott, director of research and technology development, and Robert Pappas, manager of aviation safety research and development, made presentations on the means by which the FAA and NASA coordinate in the conduct and transitioning of safety-related research. The committee's second meeting was held September 3-4, 2009, at the NASA Ames Research Center, where the committee was able to interact directly with NASA researchers from a variety of locations, including NASA Ames, Glenn, and Langley research centers. During this meeting, the committee received more detailed briefings on the research being undertaken by NASA's Aviation Safety Program from Amy Pritchett and several principal investigators.

The committee's third meeting was held on November 19-20, 2009, immediately following NASA's annual Aviation Safety Program Technical Conference. Several committee members were able to attend the conference, providing additional insight into the safety work at NASA. At this meeting, the committee heard from Jay Dryer and John Cavolowsky, directors of ARMD's Fundamental Aeronautics Program and its Airspace Systems Program, respectively. They discussed the aviation safety-related research being conducted in their ARMD programs. The committee also used this opportunity to meet with Jaiwon Shin, associate administrator for ARMD. The committee met for the final time on February 22-23, 2010, to develop this report.

The committee thanks all of the individuals who made presentations during the meetings and otherwise assisted the committee during the course of the study. Staff support for the committee's efforts was managed through the combined efforts of Thomas R. Menzies, Jr., Brian Dewhurst (through August 2009), and Paul Jackson (since August 2009). The committee wishes to give special thanks to Andrea Rebholz for her assistance with meeting arrangements and correspondence with the committee.

This report has been reviewed in draft form by individuals chosen for their diverse perspectives and technical expertise, in accordance with procedures approved by the NRC's Report Review Committee. The purpose of this independent review is to provide candid and critical comments that will assist the institution in making its published report as sound as possible and to ensure that the report meets institutional standards for objectivity, evidence, and responsiveness to the study charge. The review comments and draft manuscript remain confidential to protect the integrity of the deliberative process.

Thanks go to the following individuals for their participation in the review of this report: Anthony J. Broderick, independent aviation safety consultant; Richard J. Butler, Brigham Young University; John B. Hayhurst, the Boeing Company (retired); Roger Kasperson, George Perkins Marsh Institute at Clark University; Andrew Lacher, MITRE; John K. Lauber, Airbus SAS (retired); Robert Loewy, Georgia Institute of Technology; Najmedin Meshkati, University of Southern California; and Alfred T. Spain, JetBlue Airways Corporation (retired).

Although the reviewers listed above have provided many constructive comments and suggestions, they were not asked to endorse the conclusions or recommendations, nor did they see the final draft of the report before its release. The review of this report was overseen by Ali Mosleh, University of Maryland, College Park. Appointed by the NRC, he was responsible for making certain that an independent examination of this report was carried out in accordance with institutional procedures and that all review comments were carefully considered. Responsibility for the final content of this report rests entirely with the authoring committee and the institution.

Contents

APPENDIXES

Summary

Advancing the state of aviation safety is a central mission of the National Aeronautics and Space Administration (NASA).[1] In keeping with direction from the *National Aeronautics Research and Development Policy,*[2] issued by the Office of Science and Technology Policy, the emphasis of NASA aeronautics research is on undertaking longer-term, "foundational research" aimed at developing a stronger aviation knowledge and technology base. NASA's aeronautics research expertise and assets are also called upon to assist in finding solutions to more immediate and pressing needs of the aviation sector, particularly in the area of safety. Most responsibility for aeronautics research at NASA is housed in the Aeronautics Research Mission Directorate (ARMD). ARMD administers several major research programs, including the Aviation Safety Program, the Fundamental Aeronautics Program, and the Airspace Systems Program. While safety research is the primary domain of the Aviation Safety Program, safety-related research is undertaken in all of the ARMD programs.

Congress requested this review of NASA's aviation safety-related research programs, seeking an assessment of whether the programs have well-defined, prioritized, and appropriate research objectives; whether resources have been allocated appropriately among these objectives; whether the programs are well coordinated with the safety research programs of the Federal Aviation Administration (FAA); and whether suitable mechanisms are in place for transitioning the research results into operational technologies and procedures and certification activities in a timely manner. To undertake such an assessment requires knowledge of why NASA elected to focus its safety research on specific objectives and on certain content to meet each objective. Accordingly, the Committee for the Review of NASA's Aviation Safety-Related Programs sought from NASA an explanation of how it goes about identifying and prioritizing safety research needs. Not in a position to make its own thorough assessments of research needs and priorities, the committee refrained from providing advice to NASA on the specific research it should be doing.

The study findings with respect to each of the main aspects of the review sought by Congress are summarized next. These findings indicate that NASA's aeronautics research enterprise has made, and continues to make, valuable contributions to aviation system safety but it is falling short and needs improvement in some key respects.

[1] NASA, *2006 NASA Strategic Plan,* NP-2006-02-423-HQ, Washington, D.C., 2006, available at http://www.nasa.gov/pdf/142302main_2006_NASA_Strategic_Plan.pdf, p.13.

[2] National Science and Technology Council, *National Aeronautics Research and Development Policy,* Office of Science and Technology Policy, Executive Office of the President, Washington, D.C., December 2006, available at http://www.aeronautics.nasa.gov/releases/national_aeronautics_rd_policy_dec_2006.pdf.

KEY FINDINGS

Do NASA's Safety-Related Research Programs Have Well-Defined, Prioritized, and Appropriate Research Objectives?

Findings: *NASA needs a more objective process for prioritizing safety research.* While the objectives of ARMD's Aviation Safety Program are worthy guideposts for safety research, ARMD lacks a well-founded process for prioritizing the research needs associated with each objective, and thus for ensuring that its research is well aligned with meeting critical national aviation safety needs. *Internal interests are overemphasized in the programming of safety research.* ARMD gives undue weight to research that aligns well with its existing activities, personnel, and assets rather than the results of critical evaluations of current and emerging aviation safety needs.

The Aviation Safety Program has developed research projects to address safety issues associated with new operations, operating in hazardous conditions, loss of control, and on-board system failures and faults. The research also seeks to enhance the durability of aircraft structures and systems and to improve capabilities to analyze complex systems for safety. In the committee's view, these are worthy objectives to guide research aimed at improving both the current and the future state of aviation safety. What is not clear, however, is whether the research being undertaken in each of these areas represents the best use of ARMD's capabilities and resources to make a meaningful contribution to the targeted safety objectives.

The committee expected to find a research prioritization process that is deliberate and well informed, supported by empirical analyses, careful reviews of research being undertaken elsewhere, and advice from outside experts to enable ARMD to identify the key research needs associated with each objective and to determine where its programs can contribute the most to meeting them. The existing prioritization process, however, appears to be driven largely by ARMD's interest in employing existing personnel and assets at the NASA research centers. Those safety objectives that map well with ongoing research activities and with these internal interests are generally given priority in the programming of research and allocation of resources. By not having such a defensible, analytically based process for prioritizing its safety research, ARMD could not justify, in a convincing manner, much of the content of its research programs. Thus, in not having access to such an independent assessment of safety research needs, the committee could not determine whether ARMD's safety research programs are well prioritized to make use of available resources or identify whether changes in NASA personnel and facilities are required, and neither can ARMD.

Have Resources Been Allocated Appropriately to Research Objectives?

Finding: *Too few resources are devoted to sustaining and acquiring critical safety research capabilities.* Continued emphasis on preserving existing research expertise and assets risks degradation of ARMD's core safety research strengths and the prolonged neglect of competencies required to address new and emerging safety issues.

ARMD currently has nationally and internationally recognized research competencies in critical safety areas, such as icing research. While these existing nodes of expertise need to be recognized and their critical mass sustained, they risk being neglected as research funding is spread widely to preserve all of ARMD's research capabilities, including those that are no longer unique to NASA or of high safety relevance. Yet, even as ARMD seeks to retain and strengthen its core safety research competencies, it must give sufficient attention to investing in the research expertise that will be needed to address new and emerging safety issues, including the capability to address safety issues extending beyond aircraft to the broader aviation environment. ARMD recognizes the importance of such forward-looking investments, as evidenced by its efforts to expand and strengthen its expertise in critical software verification and validation (V&V).

To acquire the needed expertise going forward, ARMD will need to make many difficult decisions about the allocation of resources among its existing facilities and program areas, requiring that some capabilities be eliminated and others substantially scaled back. By not having in place an objective and well-informed means of assessing safety needs and priorities, building the case for such resource investments and realignments is made even more difficult.

Finding: *Too few resources and programs are devoted to stimulating innovation.* ARMD lacks the structure to elicit, explore, and develop innovative ideas to advance aviation safety.

In light of ARMD's emphasis on advanced and long-range research, the committee is surprised to find few programmatic means and resources set aside for fostering exploratory and innovative thinking on ways to solve safety problems. Even though funding such research may involve greater uncertainty about expected payoffs, taking such calculated risks may be warranted in cases where research investments to solve safety problems have reached the point of diminishing returns and for which breakthrough insights and technologies are needed. ARMD does not have any formal mechanisms to support such exploratory research and innovation.

Are the Programs Properly Coordinated with the Safety Research Programs of the FAA and Other Relevant Federal Agencies?

Finding: *Connections with the FAA, other federal agencies, and the aviation community are varied but not deep.* NASA and the FAA coordinate in the planning and conduct of safety research, and many mechanisms exist for interacting and exchanging information with other federal agencies, academia, and industry. These connections could be deepened through more inclusive and sustained reviews of NASA safety research by such outside experts.

NASA and the FAA coordinate in the planning and conduct of safety research, and many mechanisms exist for interacting and exchanging information with other federal agencies, academia, and industry. These include exemplary efforts to cooperate and collaborate in the planning and conduct of safety research, such as the work addressing high ice water content hazards in jet engines. Such varied means of coordination across federal agencies are commendable and critical, given that the interconnectedness and complexity of the aviation system demands the broadest possible cooperation and coordination of interests.

One area where greater inclusion is desirable is in seeking external reviews of ARMD safety research activities. The Aviation Safety Program activities are currently reviewed on a periodic basis by experts from the FAA and other federal agencies. Such agency reviews and consultations are vital to ensuring that the research is compatible with the work going on elsewhere in the federal government and is relevant to the operational and regulatory needs of federal agencies. Limiting these external reviews to experts from federal agencies, however, may cause ARMD to miss out on opportunities to gain additional perspective from industry and academia and to inform outside perceptions of the program's capabilities and priorities. There is a general recognition within the research community that external reviews foster higher-quality research. Yet, it is important that the reviews be conducted not only at the end stages of the research but also on a continuing basis, from the beginning of the work to its completion and transitioning.

Finding: *Internal coordination of and collaboration on safety research need improvement.* Within ARMD, there is stove-piping of research that risks system-level safety solutions not being explored and safety hazards not being addressed that arise from interactions among aviation system elements.

In the large, evolving, and complex aviation system, safety concerns can arise from interactions among system elements that may be considered safe on their own but not in combination, such as interactions among aircraft and airspace technologies and procedures. Under these circumstances, one would expect to find a significant amount of safety-related research being planned and undertaken in the other ARMD programs, often in collaboration with the Aviation Safety Program. With a few exceptions, however, ARMD's safety research programs were presented as discrete activities of the Aviation Safety Program, the Fundamental Aeronautics Program, and the Airspace Systems Program. Given the complexity and breadth of the aviation safety challenge, it is troubling to find such "stove-piping" or compartmentalization within ARMD. Even within the Aviation Safety Program itself, however, the committee observes this phenomenon, such as the relatively few collaborative activities occurring across the related objective areas of New Operations and Loss of Control and across the Aircraft Aging and Durability and Integrated Vehicle Health Management projects.

Do Suitable Mechanisms Exist for Transitioning the Research Results from the Programs into Operational Technologies and Procedures and Certification Activities in a Timely Manner?

Finding: *Demands for safety-assured technologies and procedures can conflict with NASA's emphasis on long-range, foundational research.* ARMD exploits many mechanisms to assist in furthering the technologies and procedures developed through its research; however, safety assurance and approval requirements can present vexing implementation challenges. In light of these challenges, some of ARMD's safety-related research would appear to have very limited prospects for eventual implementation—a risk that deserves more explicit consideration when ARMD programs its research.

Because much of ARMD's safety work consists of fundamental and long-term research, a high degree of operational applicability cannot be a metric for judging whether the research should be undertaken, nor can there be a strong emphasis on definitive end products and transition planning. Yet, for some of the safety-relevant research being pursued by ARMD, the resulting technologies and process are intended to support the FAA operational needs and have early application, in which case transition planning is important. The committee observes, however, that in many such cases NASA and the FAA recognize these challenges and are actively seeking to overcome them, for instance, by coordinating special research transition teams intent on identifying and addressing safety implications much earlier in the research process.

Even for research of a longer-term nature, NASA must have some understanding of eventual implementation challenges in order to judge the merits of the work. This is because achieving an accepted level of safety assurance required for certification of resulting technologies and processes may be more complicated and daunting than the development of the technologies and practices themselves. The committee believes the kind of external reviews espoused above are likely to bring such implementation challenges to the forefront, compelling early attention by those programming and engaging in the research. By applying foresight in mapping out the implementation challenges, researchers can better assess the long-term practicality of the research in relation to other research options addressing the same safety objective. Doing so will not only yield insight into these implementation challenges, but may also suggest areas where ARMD research can be helpful in overcoming them, such as in developing improved safety analysis and approval methods.

RECOMMENDATIONS

The following actions are recommended by the committee to address the shortcomings identified above.

Recommendation 1: The Aeronautics Research Mission Directorate (ARMD) should adopt a more fully informed, empirical, and documented process for identifying and prioritizing safety research needs for use in guiding its aeronautics research and development programming and investments in research expertise and capacity. A central element of this process should be the development of comprehensive aviation safety needs assessments. These assessments should be:

- Made objectively, independent of ARMD's existing research expertise, assets, and resource requirements and constraints.
- Undertaken in close coordination with the Federal Aviation Administration and other relevant federal agencies and in consultation with industry, academia, other safety-related organizations, and considering the relevant aviation safety data and literature.
- Cognizant of the safety needs being researched elsewhere in government, industry, and academia to know where critical gaps in research coverage may exist.

Recommendation 2: The Aeronautics Research Mission Directorate should establish programmatic means for encouraging more exploratory research on innovative ideas to improve aviation safety. The program should elicit and develop the promising ideas of researchers from industry, academia, other government agencies, and NASA.

Recommendation 3: The Aeronautics Research Mission Directorate's safety-related research activities should be subject to regular reviews by outside experts from the Federal Aviation Administration and other government agencies, industry, independent research institutes, and universities. These reviews, which will help in ensuring continued safety relevance, quality, implementation challenges, and successful transitioning, should be undertaken during the formative stages of the research, interim phases, and as the work is being completed. The reviews should have a prominent role in informing research programming decisions.

Recommendation 4: The Aeronautics Research Mission Directorate should develop and implement processes that will lead to more coordination and collaboration in the planning and conduct of safety research both within the Aviation Safety Program and across all its aeronautics research programs.

1

Introduction and Overview

Safety is paramount throughout the aviation enterprise. It is the central consideration in all aspects of aircraft design, manufacture, maintenance, and operations; in the training and readiness of pilots and crew; and in the design, operation, and management of the National Airspace System. As a result, aviation has a long history of technological advances accompanied by continued safety improvement. According to the National Transportation Safety Board (NTSB), U.S. mainline air carriers[1] average about 0.1 fatal accidents per million flight hours, while operators of general aviation aircraft average 10 to 15 fatal accidents per million flight hours.[2] Twenty-five years ago, fatal accident rates were three to four times higher among mainline carriers, and about one-third higher for general aviation. These marked safety improvements have occurred even as total hours flown by aircraft in the National Airspace System have more than doubled since 1985.[3] Although these safety gains represent a tremendous and hard-won accomplishment, they cannot breed complacency. Users of the nation's aviation system have come to expect more safety with each new aircraft model, innovation in air transport services, and improvement in air traffic management procedures and technologies. In a global aviation system where change is inevitable, new safety challenges must be continuously monitored, understood, and addressed.

Ensuring aviation safety is an overarching goal of the aviation industry, a priority of the federal government, and the primary responsibility of the Federal Aviation Administration (FAA). A combination of both public- and private-sector research has been, and will likely continue to be, critical to meeting these safety demands by furthering the understanding, technologies, operating procedures, and methods needed for predicting and preventing safety problems and for achieving even higher levels of safety performance. The FAA undertakes applied research in support of its operational and regulatory programs and to address pressing safety problems that arise in the field. The aviation industry also undertakes applied research in support of safe product development, use, and operations. Advancing the state of aviation safety is also a core mission of the aeronautics research and technology programs of NASA, dating back to the agency's origins as the National Advisory Committee for Aeronautics. NASA's role has tended to be longer-term in nature, aimed at advancing fundamental knowledge of aeronautics science and engineering through its research expertise and facilities.

[1] Data are for Part 121 carriers. Smaller, Part 135 carriers average about five fatal accidents per million flight hours.

[2] National Transportation Safety Board, Annual Review of Aircraft Accident Data: U.S. Air Carrier Operations, for years 2001 to 2005, retrieved from http://www.ntsb.gov/publictn/.

[3] Statistics are derived from annual NTSB accident reports retrieved from http://www.ntsb.gov/publictn/.

In recent years, NASA has placed increasing emphasis on safety research in accordance with the *National Aeronautics Research and Development Policy* (2006)[4] and the *National Plan for Aeronautics Research and Development and Related Infrastructure* (2007)[5] issued by the Office of Science and Technology Policy, the National Research Council's (NRC's) *Decadal Survey of Civil Aeronautics* (2006),[6] and a number of other advisory reports and work plans. Since 2000, aviation safety constitutes one of the main program areas in the Aeronautics Research Mission Directorate (ARMD), consisting of about 12 percent of the directorate's budget.[7] In addition, safety-related research is supported by other ARMD programs, including airspace systems and fundamental aeronautics.

STUDY CHARGE, SCOPE, AND APPROACH

Section 305 of the NASA Reauthorization Act of 2008 calls on the National Research Council to conduct an independent assessment of NASA's aviation safety-related research programs:

> The Administrator shall enter into an arrangement with the National Research Council for an independent review of the NASA's aviation safety-related research programs. The review shall assess whether
>
> (1) The programs have well-defined, prioritized, and appropriate research objectives;
> (2) The programs are properly coordinated with the safety research programs of the Federal Aviation Administration and other relevant federal agencies;
> (3) The programs have allocated appropriate resources to each of the research objectives; and
> (4) Suitable mechanisms exist for transitioning the research results from the programs into operational technologies and procedures and certification activities in a timely manner.

The committee was given this charge during its first meeting on June 22-23, 2009. During the meeting, the committee was briefed by staff of the House Committee on Science and Technology on the origins and intentions of the legislative request. The committee was also briefed by the leadership of NASA's Aviation Safety Program on the program's structure, content, funding, and management. Representatives from the FAA's aviation safety and research programs explained the research coordination and transition activities that exist between the FAA and NASA.

The committee observed that in seeking a review of safety-related research programs, Congress implied an interest that goes beyond the work of ARMD's Aviation Safety Program to include safety-related research in other programs in the directorate.

ARMD acknowledged this broader interest and proposed that the study review all work in the Aviation Safety Program as well as research having significant safety relevance in the Airspace Systems Program and the Fundamental Aeronautics Program. The committee concurred with this proposed scope, wanting to ensure that the study would cover the gamut of aviation safety-related research undertaken by NASA while recognizing that a safety interest permeates all aeronautics research.

The committee interpreted the legislative charge as seeking an assessment of whether NASA's aviation safety-related research programs are guided by a well-defined, well-prioritized, and appropriate set of objectives and whether resources are appropriately allocated among the programs. Accordingly, the committee decided against conducting a detailed, technical appraisal of the individual research activities and chose to focus instead on reviewing the means by which aviation safety-related research is prioritized, resourced, and carried out. In addition, the

[4] National Science and Technology Council, *National Aeronautics Research and Development Policy,* Office of Science and Technology Policy, Executive Office of the President, Washington, D.C., December 2006, available at http://www.aeronautics.nasa.gov/releases/national_aeronautics_rd_policy_dec_2006.pdf.

[5] National Science and Technology Council, *National Plan for Aeronautics Research and Development and Related Infrastructure,* Office of Science and Technology Policy, Executive Office of the President, Washington, D.C., December 2007, available at http://www.aeronautics.nasa.gov/releases/aero_rd_plan_final_21_dec_2007.pdf.

[6] National Research Council, *Decadal Survey of Civil Aeronautics: Foundation for the Future,* The National Academies Press, Washington, D.C., 2006.

[7] Amy Pritchett, Director, NASA Aviation Safety Program, "Safety-Related Research in NASA's Aeronautics Research Mission Directorate: Overview," presentation to the committee, September 3, 2009, p. 17.

committee would examine how well the research is coordinated with the safety programs of the FAA and whether adequate means exist for transitioning the research to application where appropriate.

Concerned about the limited time available for this study, the committee asked NASA to identify the objectives guiding its safety research and to address the following set of questions:

- What drove the selection of the research objectives and why did NASA elect to pursue them?
- What research activities are supporting each objective and how are the activities being conducted?
- How are the research activities coordinated with the FAA and other federal government agencies?
- What resources have been allocated to each objective?
- What are the plans for transitioning the research results?
- Do the activities in support of the objectives have the appropriate people and expertise?

The committee convened for a second time at NASA's Ames Research Center on September 3-4, 2009. At this meeting, the committee received more detailed briefings on the Aviation Safety Program and its objectives, as well as responses to the series of questions posed above. The committee used this opportunity to meet with researchers managing and working on specific aviation safety projects and to observe research results. During its third meeting on November 19-20, 2009, the committee met with the leadership of the Airspace Systems Program and the Fundamental Aeronautics Program to learn more about the safety-related work in their programs. The committee also met with the associate administrator for ARMD, seeking a better understanding of how safety research is programmed across the directorate. During this third meeting, several members of the committee also attended the Aviation Safety Technical Conference sponsored by the Aviation Safety Program, providing additional exposure to the work being undertaken (see Appendix C for conference program content). The committee met for the final time on February 22-23, 2010, to develop this report.

The information and insights gleaned during these meetings were central to the committee's undertaking of this review and fulfilling its charge.

ORGANIZATION AND CONTENT OF NASA'S AVIATION SAFETY-RELATED RESEARCH

NASA manages four mission directorates: Science, Exploration Systems, Space Operations, and Aeronautics Research. ARMD accounts for about 2.5 percent of NASA's fiscal year (FY) 2010 budget.[8] ARMD's research goals are to improve airspace capacity and mobility, aviation safety, and aircraft performance while reducing noise, emissions, and fuel burn.[9]

ARMD manages three major research programs that support the safety goal: the Aviation Safety Program, the Airspace Systems Program, and the Fundamental Aeronautics Program. Although most of the agency's research directly related to aviation safety is undertaken by the first, the latter two also undertake research with direct safety relevance. Each of these three ARMD programs is charged with conducting "long-term, cutting-edge research for the benefit of the broad aeronautics community."[10] The stated aim of the Aviation Safety Program is to develop innovative concepts, tools, and technologies to improve the intrinsic safety attributes of current and future aircraft.[11] The Airspace Systems Program aims to develop revolutionary concepts, capabilities, and technologies that will enable significant increases in the capacity, efficiency, and flexibility of the National Airspace System. The Fundamental Aeronautics Program seeks to enable revolutionary changes for vehicles that fly in all speed regimes.

[8] See "A New Era of Responsibility: Renewing America's Promise. The National Aeronautics and Space Administration 2010 Budget," information prepared by the White House Office of Management and Budget, Washington, D.C., available at http://www.nasa.gov/pdf/315067main_fy10_nasa.pdf.

[9] See NASA's Aeronautics Research Mission Directorate Web site at http://www.aeronautics.nasa.gov/.

[10] See NASA, *FY 2009 Performance and Accountability Report,* NP-2009-11-633-HQ, NASA, Washington, D.C., available at http://www.nasa.gov/pdf/403618main_NASA_FY09_Performance_Accountability_Report.pdf.

[11] A smaller, fourth research program, the Integrated Systems Research Program, is a relatively new effort whose goal is to conduct integrated-systems research on promising aeronautical concepts and technologies, exploring, assessing, and demonstrating their benefits in a relevant environment.

TABLE 1.1 Aeronautics Research Mission Directorate 5-Year Budget, FY 2010 to FY 2014

	Funding (millions of dollars)				
	FY 2010	FY 2011	FY 2012	FY 2013	FY 2014
Aviation Safety Program	60.1	59.6	59.2	61.7	62.5
Aircraft Aging and Durability	11.4	11.2	11.7	12.1	12.1
Integrated Intelligent Flight Deck	12.5	13.3	11.6	12.6	13.3
Integrated Resilient Aircraft Control	16.4	17.0	17.6	18.2	18.2
Integrated Vehicle Health Management	19.8	18.2	18.3	18.9	18.9
Airspace Systems Program	81.4	82.9	83.9	87.2	88.3
Fundamental Aeronautics Program	228.4	230.0	233.6	239.0	245.9
Aeronautics Test	74.7	77.1	77.2	76.6	78.7
Integrated Systems	62.4	64.4	67.1	64.4	60.5
Aeronautics Total	507.0	514.0	521.0	529.0	536.0

SOURCE: NASA, ARMD FY 2010 Budget Proposal, available at http://www.nasa.gov/pdf/345954main_7_ Aeronautics_%20FY_2010_UPDATED_final.pdf.

ARMD's Aviation Safety Program

The Aviation Safety Program was formed in 2000, shortly after the White House Commission on Aviation Safety and Security[12] (known as the Gore Commission) challenged the federal government and the aviation industry to reduce the aviation accident rate by 80 percent. As stated in NASA's 2003 Strategic Plan, the program's purpose was "to develop prevention, intervention, and mitigation technologies and strategies aimed at one or more causal, contributory, or circumstantial factors associated with aviation accidents,"[13] commensurate with the Gore Commission's recommendations to achieve near-term reductions in accident and fatality rates. In 2006, ARMD reorganized its research program coincidental with the 2006 *National Aeronautics Research and Development Policy,* which stressed the importance of NASA aligning its research with the needs of the Next Generation Air Transportation System (NextGen) and maintaining a broad foundational research program aimed at "preserving the nation's intellectual stewardship and mastery of aeronautics core competencies."[14]

The Aviation Safety Program's funds are divided among four major research projects, or portfolios: Aircraft Aging and Durability (AAD), Integrated Intelligent Flight Deck (IIFD), Integrated Resilient Aircraft Control (IRAC), and Integrated Vehicle Health Management (IVHM). The approved and proposed budgets for the programs for FY 2010 to FY 2014 are shown in Table 1.1. The program is managed by a program director and a deputy director located at NASA headquarters in Washington, D.C., while each of the four projects is led by a principal investigator, a project manager, and a project scientist working from the NASA research centers at Langley, Ames, Glenn, or Dryden. Most of the research is performed by scientists and engineers at the centers through research agreements among project leaders and center branch chiefs. To a more limited extent, projects also engage federal government agencies, the aviation industry, and academia.

The general subject matter in each the four projects of the Aviation Safety Program are described next and summarized in Box 1.1.

[12] See Executive Order 13015: White House Commission on Aviation Safety and Security, August 22, 1996, Federal Register Document 96-21996, available at http://ntl.bts.gov/DOCS/eo13015.html.

[13] See NASA, *2003 Strategic Plan,* NP-2003-01-298-HQ, Washington, D.C., available at http://www.nasa.gov/ pdf/1968main_strategi.pdf.

[14] National Science and Technology Council, *National Aeronautics Research and Development Policy,* Office of Science and Technology Policy, Executive Office of the President, Washington, D.C., December 2006, available at http://www.aeronautics.nasa.gov/releases/national_aeronautics_rd_policy_dec_2006.pdf.

BOX 1.1
Aims of Aviation Safety Program Research Portfolios

Aircraft Aging and Durability

- Improving understanding of how cracks in materials start and grow;
- Developing computer models to predict crack growth in metallic and nonmetallic materials;
- Identifying the durability of nonmetallic (composite) materials used for aircraft fuselages;
- Improving the ability to detect bonded joint degradation;
- Identifying the long-term service and environmental needs of composite jet engine containment cases;
- Improving understanding of how quickly engine disks operating at hotter temperatures degrade over time;
- Developing new software tools to better identify and repair wiring faults.

Integrated Intelligent Flight Deck

- Assigning clear roles and responsibilities to human and automated agents;
- Predicting human and automated agent performance in both normal and abnormal conditions;
- Evaluating human, automation, and joint human-automation performance to help make automation more comprehensible to pilots;
- Predicting joint human-automation performance in operating environments that are not yet realized, such as NextGen's trajectory-based operations;
- Achieving a "better than visual" flight operations capability;
- Enabling a highly collaborative working environment for flight deck system operators.

Integrated Resilient Aircraft Control

- Understanding the dynamics of current and future aircraft when in damaged and upset conditions;
- Developing control systems that adapt reliably to both the anticipated and the unanticipated;
- Developing aircraft guidance for emergency operation;
- Modeling and sensing airframe and engine icing;
- Modeling effective and reliable human-automation systems.

Integrated Vehicle Health Management

- Developing on-board systems that can predict, detect, diagnose and propose solutions for failures that involve the airframe, propulsion systems, avionics (hardware) and software;
- Creating reliable and accurate systems that reveal vehicle or airspace problems before they become accidents;
- Designing and testing new sensors that detect and display airframe and engine icing and other environmental hazards.

SOURCE: Aviation Safety Program fact sheet, available at http://www.aeronautics.nasa.gov/pdf/avsafe_fs.pdf.

Aircraft Aging and Durability

Out of concern that most previous research on aircraft aging and durability has been largely reactive in nature and based more on observations than on fundamental understanding, NASA's stated goal for its AAD portfolio is to perform foundational research in aging science that will yield multidisciplinary analysis and optimization capabilities for the detection, prediction, mitigation, and management of aging-related hazards for future civilian and military aircraft.[15] The focus of AAD research, therefore, is on aging and damage processes in "young" aircraft, rather than life extension of legacy vehicles. The emphasis of the project's research is on new and emerging material systems and fabrication techniques and the potential hazards associated with aging-related degradation. Stated goals are to take a proactive approach to identify aging-related hazards before they become critical and to develop technology and processes to incorporate aging mitigation and maintenance into the design of future aircraft.[16]

Integrated Intelligent Flight Deck

NASA anticipates that methods for piloting aircraft will change dramatically over the coming decades with the transition to NextGen. NASA envisions changes that will lead to aircraft and airspace systems that are more complex, along with greater complexity of flight deck systems and procedures. The stated goal of IIFD, therefore, is to develop tools, methods, principles, guidelines, and technologies for the advent of revolutionary flight deck systems that ensure safe operations.[17] In so doing, IIFD seeks to enhance the ability to predict demands and create a comprehensive set of capabilities (e.g., technologies, procedures, and specifications for crew training) for meeting the demands of the kinds of operational concepts proposed for NextGen. To this end, the program is seeking to develop both predictive and generalizable methods and models for designing technologies and operating procedures suitable for use by the aviation community for systematically considering human and technology performance throughout procedure and technology design. IIFD also seeks revolutionary advancements in the capability and performance of avionics technology in selected areas where new demands for high-integrity capabilities are required, such as external hazard detection.

Integrated Vehicle Health Management

The stated goal of IVHM is to develop validated tools, technologies, and techniques for automated detection, diagnosis, and prognosis that enable mitigation of adverse events during flight, such as events that arise from damage, degradation, and environmental hazards.[18] One of the research challenges of IVHM is to further the development of real-time automated reasoning and decision-making tools and techniques to integrate messages from the health management systems of individual aircraft and combine them with results from fleet-wide vehicle health assessments. Although the project title implies an aircraft focus, its scope is intended to be broader by encompassing events at the air transportation system level. For instance, IVHM research projects are aimed at developing probabilistic models of fault and failure modes and data-mining algorithms to analyze data sources from current aircraft fleets to develop models of potential system failures. Through development of data-mining capabilities for system-wide data sets, the project seeks to further the development of analytic techniques for identifying precursors to failures.

---

[15] NASA, *Aviation Safety Program Aircraft Aging and Durability Project Technical Plan Summary,* Washington, D.C., available at http://www.aeronautics.nasa.gov/nra_pdf/aad_technical_plan_c1.pdf.

[16] NASA, *Aviation Safety Program Aircraft Aging and Durability Project Technical Plan Summary,* Washington, D.C., available at http://www.aeronautics.nasa.gov/nra_pdf/aad_technical_plan_c1.pdf.

[17] NASA, *Integrated Intelligent Flight Deck Technologies: Technical Plan Summary (FY2009-FY2013),* Aviation Safety Program, Aeronautics Research Mission Directorate, NASA, Washington, D.C., March 13, 2009, available at http://www.aeronautics.nasa.gov/nra_pdf/iifd_tech_plan_2009.pdf.

[18] See NASA, *Integrated Vehicle Health Management Technical Plan, Version 2.03,* Aviation Safety Program, Aeronautics Research Mission Directorate, NASA, Washington, D.C., November 2, 2009, available at http://www.aeronautics.nasa.gov/nra_pdf/ivhm_tech_plan_c1.pdf.

Integrated Resilient Aircraft Control

IRAC conducts research to advance the state of aircraft flight control to provide on-board control resilience for ensuring stability, maneuverability, and safe-landing capabilities in the presence of adverse conditions (for example, a pilot's loss of control of an aircraft caused by environmental factors, actuator and sensor faults or failures). The stated goal of the project is to arrive at a set of validated multidisciplinary integrated aircraft control design tools and techniques for enabling safe flight in the presence of such adversities.[19] By advancing understanding of the dynamics involved in loss of control, the project seeks to gain a better understanding of how an adaptive system can best regain control. The focus of the effort is on current and next generation subsonic civil air transports, although the program technical plan states that a majority of the challenges addressed are general in nature and are thus applicable to a larger class of aviation vehicles. IRAC research results are being validated through the use of NASA simulators, wind tunnels, and sub- and full-scale flight test vehicles.

Safety-Related Research in Other ARMD Programs

The Airspace Systems Program's goal is to develop and demonstrate future concepts, capabilities, and technologies that maintain safety and meet NextGen requirements to enable major increases in air traffic management capacity, flexibility, and efficiency. While the Airspace Systems Program must consider safety across its research portfolio, some of its projects aimed primarily at increasing system capacity and efficiency are especially relevant to safety. For example, runway overruns are one of the top safety issues in aviation. Through its program of research on automated separation assurance for en route and transition airspace and the airport surface, the Airspace Systems Program seeks to develop and evaluate airport traffic conflict-detection algorithms, resolution advisories, and alerting display concepts. Its wake vortex research program seeks to safely increase runway capacity by improving existing modeling capabilities, developing a better understanding of wake measurement accuracy, and developing a probabilistic model to enable a dynamic aircraft separation capability. The Airspace Systems Program is also examining methods to define weather impacted areas that can be integrated with air traffic management tools.

The Airspace Systems Program collaborates with IIFD researchers to evaluate the impact of communications alternatives (datalink vs. voice) and advanced displays on flight-deck workload and situational awareness. In addition, the Joint Program Development Office's (JPDO) integrated workplan calls for NASA to take the lead in developing the capability to perform complex systems validation and verification (V&V) for NextGen.[20] The Aviation Safety Program and the Airspace Systems Program share the responsibility for furthering V&V modeling techniques for safety-critical concepts and technologies.

The main goal of the Fundamental Aeronautics Program is to develop capabilities necessary to address national challenges in air transportation, including noise, emissions, fuel consumption, acceptable supersonic flight over land, mobility, and the ability to ascend or descend through planetary atmospheres. While safety considerations are inherent in much of the research conducted within the Fundamental Aeronautics Program, the focus of the research is largely on performance, efficiency, and environmental challenges for future air vehicles. Nevertheless, the program does coordinate with the Aviation Safety Program as appropriate to help ensure that certain safety research concerns are addressed.

Two research areas within the Fundamental Aeronautics Program's Subsonic Rotary Wing Project are focused on rotorcraft safety, specifically flights involving hazardous icing conditions and crash survivability. Out of concern that icing is a barrier to all-weather operations for rotorcraft in NextGen, the program is seeking to improve the agency's existing icing tools, methods, and databases to make them applicable to rotorcraft application for design and certification. The work is being conducted in the Icing Research Branch at Glenn Research Center, which has a long history of examining rotorcraft icing research. Likewise, believing that rotary wing configurations may

---

[19] See NASA, *Integrated Resilient Aircraft Control Technical Plan,* Aviation Safety Program, Aeronautics Research Mission Directorate, NASA, Washington, D.C., May 1, 2009, available at http://www.aeronautics.nasa.gov/nra_pdf/irac_tech_plan_c1.pdf.

[20] See Joint Planning and Development Office (JPDO), *NextGen Integrated Work Plan Version 1.0,* posted on September 30, 2008, at http://www.jpdo.gov/iwp.asp. JPDO was established by the 105th Congress as part of the Vision 100—Century of Aviation Reauthorization Act (Public Law 108-176).

have application for large-scale passenger transportation in NextGen, NASA is seeking to improve the crash and survivability characteristics of these aircraft, which have unique crashworthiness and survivability requirements due to their configurations (for example, heavy mechanical components located above the passenger compartment) and flight operations (for example, impact velocities that often contain a significant vertical component). Current practice for demonstrating crashworthiness relies mainly on full-scale crash testing, the cost of which limits the amount of data available to establish confidence in designs and validates designs only for a specific set of crash parameters and terrain. The Fundamental Aeronautics Program's research seeks to improve the models and methodologies (validated by performing component and full-scale helicopter crash tests) used to predict crashworthiness of rotorcraft and to develop and demonstrate advanced structural concepts for improved energy absorption and crashworthiness. This research employs NASA's specialized expertise and facilities in crashworthiness technologies, including its Landing and Impact Research Facility and scientific and engineering expertise in impact dynamics for spacecraft return.

REPORT ORGANIZATION

The remainder of this report consists of three chapters. Chapter 2 explains the sources of input in NASA's determination of important safety research needs and how the agency uses this input to establish research priorities. Chapter 3 examines the research that is being undertaken to address each of six safety concerns that NASA presented as being the main objectives of safety research. These objectives are assessed with respect to each of the four questions in the legislative request for this study. Chapter 4 integrates the information from these chapters to produce a series of findings and concludes with recommended actions.

2

NASA's Aviation Safety Research Prioritization

The first task in the request for this study calls for an assessment of whether NASA's safety-related research programs have well-defined, prioritized, and appropriate research objectives. The committee interpreted this task as a congressional interest in knowing whether NASA has well-founded objectives to guide its aviation safety research. Addressing this task thus requires an understanding of what these objectives are, how they were established and prioritized, and how they are being used.

At the highest level, NASA's aviation safety research is guided by the agency's overarching mission for aeronautics research. This mission has changed in recent years, particularly in the safety domain. Having significant influence over the current mission is the *National Aeronautics Research and Development Policy* (2006)[1] and the *National Plan for Aeronautics Research and Development and Related Infrastructure* (2007),[2] both issued by the Office of Science and Technology Policy. The first, referred to as the "National Policy," assigns roles to each federal agency engaged in aeronautics research and development (R&D), defining the timeframe and breadth of their research efforts. The second, referred to as the "National Plan," identifies a series of near- to long-term aviation safety challenges that the FAA, NASA, and other federal government agencies are expected to address commensurate with their research roles.

NASA's aviation safety research mission has evolved over the past dozen years in response to changing policy guidance, from the 1997 report by the White House Commission on Aviation Safety and Security to the more recent National Policy. The National Plan sets forth fundamental safety challenges intended to inform and guide NASA's aviation safety research, as well as that of the FAA and other federal government agencies. In addition to these national policies and plans, NASA aviation safety research programs receive input and guidance from a number of other sources, including the National Research Council's (NRC's) 2006 *Decadal Survey of Civil Aeronautics*,[3] the Joint Program Development Office (JPDO), and the Commercial Aviation Safety Team (CAST). The committee

---

[1] National Science and Technology Council, *National Aeronautics Research and Development Policy,* Office of Science and Technology Policy, Executive Office of the President, Washington, D.C., December 2006, available at http://www.aeronautics.nasa.gov/releases/national_aeronautics_rd_policy_dec_2006.pdf.

[2] National Science and Technology Council, *National Plan for Aeronautics Research and Development and Related Infrastructure,* Office of Science and Technology Policy, Executive Office of the President, Washington, D.C., December 2007, available at http://www.aeronautics.nasa.gov/releases/aero_rd_plan_final_21_dec_2007.pdf.

[3] National Research Council, *Decadal Survey of Civil Aeronautics: Foundation for the Future,* The National Academies Press, Washington, D.C., 2006.

examined and assessed all these sources, as well as the process employed by NASA to prioritize safety needs for the programming and resourcing of its safety research programs.

NASA'S AVIATION SAFETY RESEARCH ROLE AND MISSION

NASA has an important role in aeronautics research that dates back to its predecessor agency, the National Advisory Committee for Aeronautics (NACA).[4] In particular, early NACA research and test facilities were instrumental in advancing the safety of civil aviation. Through research conducted over more than four decades—from pioneering tests during the 1920s of wing and propeller icing in refrigerated wind tunnels to the building of design data needed to achieve the stability and control characteristics essential for the introduction of passenger jets in the 1950s—NACA was a leading contributor to aviation safety. NASA's subsequent programs of research and testing in a variety of areas, such as airborne wind shear detection, microwave landing systems, and head-up displays, have helped advance the safety performance of each generation of aircraft and its operating environment. At the same time, NASA has also played a critical role in addressing pressing safety issues that arise from incidents, from its work in the 1960s on means to prevent bird strikes to its collaborations with the FAA in the 1980s to develop a better understanding of the causes of and corrective actions for wind shear.

Continued growth in aviation has generated significant additional challenges to aviation, from community noise and emissions to airline delays and security threats. These challenges have tapped significant R&D resources from NASA, the FAA, other federal government agencies, and the aviation industry. As the demand on research has risen, so too has the demand for results from safety research that can be applied to immediate safety problems and concerns. The 1996 crashes of TWA Flight 800 and Valujet Flight 592 prompted the White House to convene the Commission on Aviation Safety and Security, chaired by Vice President Al Gore (the Gore Commission). This commission was charged with recommending ways to improve aviation safety and security through changes in procedures, regulation, and research and technology. The 1997 report recommended that a principal focus of government and industry safety efforts should be on reducing the rate of commercial airline accidents by a factor of five within a decade. To do so, the report stressed the importance of partnerships, and specifically urged NASA to expand its collaboration with the FAA and the aviation industry to improve airline safety.

In keeping with the Gore Commission's advice, the FAA and NASA signed an agreement in 1999 formalizing the agencies' mutual commitment to developing technologies with the greatest potential for reducing the commercial aviation accident rate.[5] The two agencies agreed to engage in joint research in a number of areas, such as aging aircraft, wake vortex research, wind shear prediction, and aircraft icing detection.[6] In addition, NASA emphasized its role in bringing about early improvements in commercial aviation safety. It set a target of reducing the airline fatal accident rate by 50 percent (when compared with baseline levels from 1990 to 1996),[7] as well as a 10-year goal for an 80 percent reduction.

Five years into its realigned safety research program, however, NASA found itself struggling to measure progress in achieving the airline safety targets. The performance metrics of airline accident and fatality rates proved to be problematic for gauging the impact of research over a relatively short time horizon. Hence, by 2005 NASA was intent on reorienting its safety research program to deemphasize the connection to current accident rates. The policy guidance in the 2006 National Policy coincided with NASA's decision to change course. The National Policy espoused a more forward-looking aeronautics research program, one that would make better use of each federal agency's core competencies and unique research capabilities for the aviation community generally. It continued to

[4] See Roger E. Bilstein, *Orders of Magnitude: A History of the NACA and NASA, 1915-1990*, NASA History Series, NASA SP-4406, NASA, Washington, D.C., 1989.

[5] Memorandum of Understanding between Federal Aviation Administration (FAA) and the National Aeronautics and Space Administration (NASA) concerning Aviation Safety Research, FNA 08, July 2, 1999.

[6] See NASA, *Fiscal Year 2003 Performance and Accountability Report,* Washington, D.C., 2004, available at http://www.nasa.gov/pdf/56091main_NASA_FY2003_PAR.pdf.

[7] See NASA, *Strategic Plan 2000,* NPD 1000.1b, Office of Policy and Plans, Code Z, NASA, Washington, D.C., September 2000, available at http://www.hq.nasa.gov/office/codez/plans/pl2000.pdf, and NASA, *Aerospace Technology Enterprise Strategic Plan,* Code R, NASA, Washington, D.C., April 2001, available at http://www.hq.nasa.gov/office/codez/plans/AST00plan.pdf.

stress the importance of research partnerships, but emphasized a stronger federal role in conducting longer-term, *foundational* research consisting of basic research and developing a strong aeronautics knowledge and technology base to overcome challenges to technological progress. NASA, in particular, was charged with maintaining such a foundational research effort. The National Policy also reflected a new federal emphasis on bringing about the Next Generation Air Transportation System (NextGen) as a means of increasing system capacity and reliability while improving safety and security. It therefore stressed the importance of NASA's engagement in longer-range, fundamental research to address the needs of NextGen.

Commencing in 2005, NASA's comprehensive restructuring of its aeronautics research programs into the Aeronautics Research Mission Directorate (ARMD) aligned well with the direction of the National Policy. ARMD's declared focus was on pursuing "long-term, cutting-edge research that expands the boundaries of aeronautical knowledge for the benefit of the broad aeronautics community."[8] Consistent with this emphasis, NASA updated its strategic plan in 2006 establishing an aeronautics research goal to "advance knowledge in the fundamental disciplines of aeronautics, and develop technologies for safer aircraft and higher capacity airspace systems."[9] The strategic plan called for research that would "identify and develop tools, methods, and technologies for improving overall aircraft safety of new and legacy vehicles operating in the Next Generation Air Transportation System."[10] Accordingly, ARMD chose a safety research approach that would be more prognostic, aimed at predicting and preventing safety problems rather than reacting to incidents. Particular emphasis was placed on understanding the safety implications of future aircraft operating in NextGen.

During 2005 and 2006, ARMD worked toward aligning its research programs with this new program emphasis. As explained in Chapter 1, the Fundamental Aeronautics Program was specifically charged with conducting cutting-edge research that produces concepts, tools, and technologies that enable the design of vehicles that fly through any atmosphere at any speed. The Airspace Systems Program was charged with addressing the fundamental air traffic management research needs of NextGen and striving to develop revolutionary concepts, capabilities, and technologies to enable significant increases in the capacity, efficiency, and flexibility of the National Airspace System. The Aviation Safety Program was charged with focusing on developing revolutionary tools, methods, and technologies to improve the inherent safety attributes of current and future aircraft that will be operating in the evolving National Airspace System.

ACTIVITIES AND ENTITIES THAT INFORM NASA'S SAFETY RESEARCH

Developed with assistance and leadership from NASA, the FAA, and other federal government agencies, the National Plan lays out the research roles and priorities of these agencies in the context of a transformed, NextGen aviation system. With respect to safety, it points to the potential for new and diverse aircraft operations, including new general aviation, advanced rotorcraft, very light jets, and unmanned aircraft,[11] that may present new and complex safety challenges that will require research and development of new safety technologies and operating procedures. The National Plan also acknowledges a continued need for research to address existing safety issues but observes that "the current system has reached a state where low accident levels for commercial aviation, and the traditional forensic investigation approach to aviation safety, are yielding fewer insights capable of significantly improving aviation safety."[12] Consistent with NASA's revised aviation safety program, the National Plan stresses the importance of research to advance preventative and prognostic techniques.

The National Plan identifies a series of "fundamental research challenges" that include the following three overarching safety goals and seven safety challenges:

[8] NASA, *FY 2006 Performance and Accountability Report*, available at http://www.nasa.gov/pdf/ 167682main_ FY_2006_NASA_PAR_508.pdf, p. 34.

[9] See NASA, *2006 NASA Strategic Plan,* NP-2006-02-423-HQ, NASA, Washington, D.C., 2006, available at http://www.nasa.gov/pdf/ 142302main_2006_NASA_Strategic_Plan.pdf, p. 13.

[10] NASA, *2006 NASA Strategic Plan*, 2006, p. 42.

[11] Technical Appendix to the *National Plan for Aeronautics Research and Development and Related Infrastructure,* available at http://www.whitehouse.gov/files/documents/ostp/default-file/technical_appendix_ high.pdf, p. 61.

[12] NTSC, *National Plan for Aeronautics Research and Development*, 2007, p. 19.

Goal 1: Safer Vehicle Designs, Structures, and Subsystems

- Monitoring and assessing aircraft health at both the material and component level.
- Rapidly and safely incorporating advances in avionics.
- Stabilizing and maneuvering next-generation aircraft in response to safety issues in the NextGen airspace.

Goal 2: Safer Vehicle Air and Ground Operations

- Understanding and predicting systemwide safety concerns of the airspace system and the vehicles envisioned by NextGen.
- Understanding the key parameters of human performance in aviation.
- Ensuring safe operations for the complex mix of vehicles anticipated within the next-generation airspace.

Goal 3: Enhanced Crash Survivability

- Enhancing the probability that passengers and crew will survive crash impact and escape safely when accidents do occur.

These challenges are characterized as being top priorities, intended to provide high-level guidance for foundational, advanced aircraft systems, and air transportation management systems R&D through 2020.

Although the National Plan does not explain how these specific safety challenges were identified, it does point to safety-related research needs that have been identified by other sources. Over the past several years, a number of other reports and plans have been issued to advise and guide NASA's aeronautics research program, both generally and with respect to safety research. In 2003, Congress created the JPDO to guide and coordinate the efforts of NASA, the FAA, the Department of Defense, the aviation industry, and other entities responsible for bringing about NextGen, and in 2006, NASA commissioned the *Decadal Survey of Civil Aeronautics* from the NRC to provide guidelines for investment in aeronautics research and technology (R&T) development. Other inputs include the safety data-driven analyses of the CAST administered in conjunction with the International Civil Aviation Organization (ICAO).

The purpose of the NRC's *Decadal Survey* was to inform a longer-term strategy for NASA's involvement in civil aeronautics research. It was undertaken by a 15-member steering committee informed by the work of five panels with expertise in specific research disciplines, such as propulsion, materials, control, and autonomous systems. The expert panels focused on identifying and prioritizing individual aeronautics research and technology (R&T) challenges that are particularly suited to NASA aeronautics research. All of the 39 highest-rated challenges pertaining to safety (scoring a "9" for safety and reliability) are listed in Appendix D. Discipline areas having the greatest number of high-priority safety challenges were (1) Intelligent and Autonomous Systems, Operations and Decision Making, Human Integrated Systems, and Networking and Communications (15 challenges) and (2) Dynamics, Navigation, Control, and Avionics (13 challenges).

As a partner in the JPDO, NASA participates in JPDO working groups and assists in the development of its relevant safety-related R&D plans. In September 2008, the JPDO released its *Next Generation Air Transportation System Integrated Work Plan.*[13] The Integrated Work Plan (IWP) is viewed by JPDO partners as a tool to build commitments and coordinate efforts and resources to help bring about NextGen. As such, it outlines federal government and industry responsibilities and collaborations, proposing a series of milestones and timelines for actions, including R&D activities. The IWP calls on NASA to conduct R&D in a number of safety-related areas, including advances in analytic tools for safety assurance and risk management, runway incursion prevention and detection systems, and new aircraft materials and designs that achieve long-term air-

[13] Joint Planning and Development Office (JPDO), *Next Generation Air Transportation System Integrated Work Plan Version 1.0,* posted on September 30, 2008, at http://www.jpdo.gov/iwp.asp. JPDO was established by the 105th Congress as part of the Vision 100—Century of Aviation Reauthorization Act (Public Law 108-176).

worthiness. In particular, the IWP points to the following safety-related areas requiring coordinated research:

- Tools to support the NextGen Aviation Safety Information Analysis and Sharing (ASIAS) capability;
- Methods for verification and validation of complex systems to support NextGen risk assessment and certification decisions; and
- Performance models that capture human variability and error in highly automated NextGen systems to aid in the development of risk-reducing interfaces, procedures, and training.

NASA's Aviation Safety Program and the Airspace Systems Program are listed in the IWP as having sole or shared primary responsibility for an array of NextGen-related research, many examples of which are shown in Box 2.1.

NASA connects with the aviation industry through CAST, which employs a data-driven strategy to develop and promote government and industry safety initiatives aimed at reducing the risk of commercial aviation fatalities. Composed of industry and government safety experts, CAST analyzes reports and data from aircraft incidents

BOX 2.1
Examples of NASA Safety-Related Research Roles in the Joint Planning and Development Office's Integrated Workplan

- Vulnerability discovery to support analysis tool development for safety assurance and safety risk management (SRM) using the Federal Aviation Administration's Aviation Safety Information Analysis and Sharing (ASIAS) capabilities
- Operator situational awareness for low-visibility terminal and airport surface operations
- Air and ground separation management architectures satisfying NextGen's higher capacity and safety requirements
- Air- and ground-based runway incursion detection technologies
- Taxi operations in low-visibility conditions
- Cockpit information requirements and procedures for independent parallel and converging runway approaches in low visibility
- Low-visibility dependent multiple approach procedures
- New materials and advanced aircraft designs supporting long-term aircraft structural airworthiness
- System health management to support NextGen equipage decisions
- Adaptive control systems that support the prevention and recovery from upset conditions and that adapt and respond to rapidly changing conditions
- Methods and algorithms to support the verification and validation of complex systems
- System risk assessment and management models addressing functional allocation across flight operator and air navigation service providers
- Risk-reducing systems interfaces, procedures, and training
- Contributing-factor analysis and fault management to support analysis tool development for safety assurance and SRM using ASIAS capabilities
- Operating procedures for human weather forecasters using automated systems
- Situational awareness technologies in low-visibility and surface operations
- Vulnerability detection tools that will monitor and analyze safety information environments and data resources

SOURCE: Joint Planning and Development Office, *NextGen Integrated Work Plan Version 1.0,* posted on September 30, 2008, at http://www.jpdo.gov/iwp.asp.

worldwide. Teams of experts from member organizations are then tasked with developing methods to more fully understand the chain of events leading to incidents and to identify solutions. CAST aided the FAA in the creation of the agency's ASIAS program. NASA is a member of CAST, and thus managers from the Aviation Safety Program participate in CAST's Joint Implementation Measurement Data Analysis Team (JIMDAT), which develops a master safety plan, measures effectiveness, and identifies future areas of study, all key inputs to the CAST Safety Plan. The director of NASA's Aviation Safety Program is also a member of the Executive Board of ASIAS and thus approves ASIAS-directed safety analyses. CAST has recently expanded its focus to include identifying new and emerging safety risks to enable the implementation of proactive measures before new types of accidents occur. NASA is also participating in this initiative, referred to as "FAST"—Future Aviation Safety Team.

NASA'S MEANS OF RESEARCH PRIORITIZATION

Aware of these many external sources of input, the committee asked NASA to describe how it uses this information, and any other sources, to establish aviation-safety research objectives and prioritize them for the purpose of programming and resourcing research. Figure 2.1 depicts the process as described schematically.

The committee was told that a major function of the Aviation Safety Program office is to review its existing research portfolio and any proposed research for consistency with NASA's mission, the fundamental safety challenges identified in the National Plan, and identifiable safety research needs. To identify safety research needs,

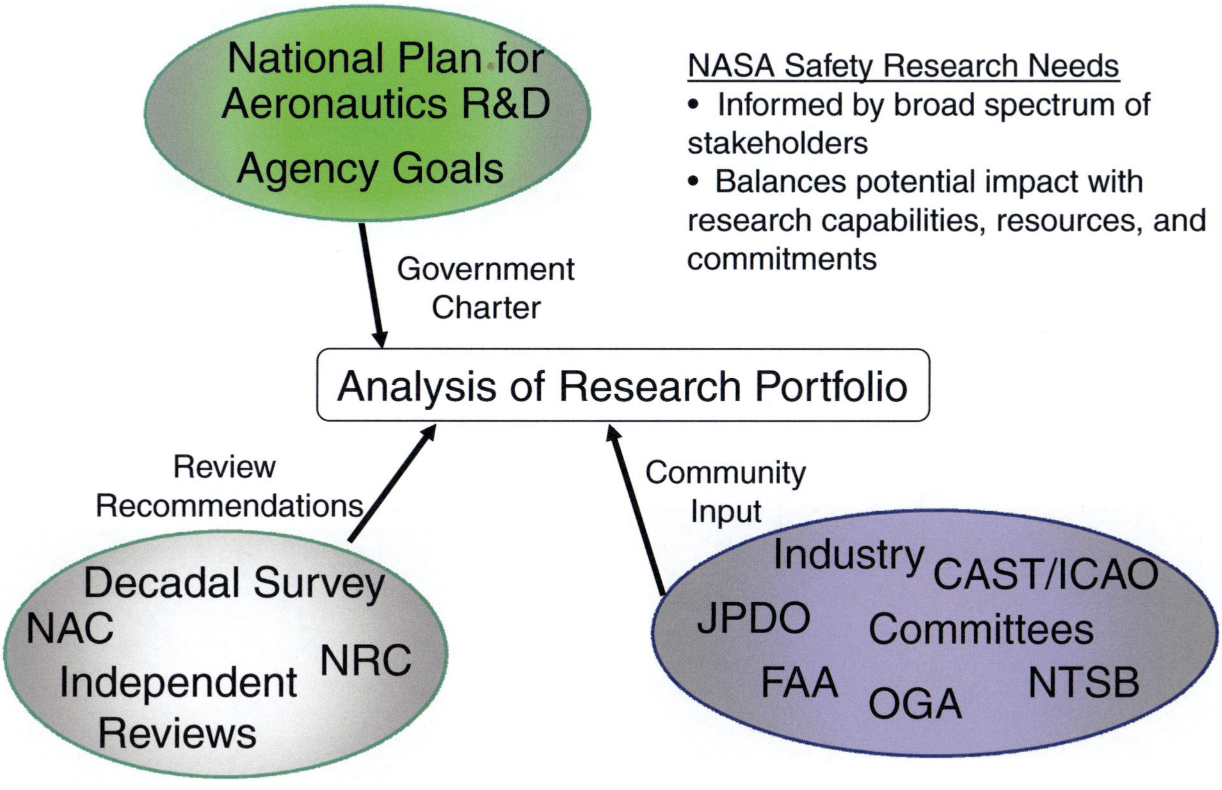

FIGURE 2.1 NASA depiction of its process for analyzing its safety research portfolio in relation to research needs. SOURCE: Amy Pritchett, Director, NASA Aviation Safety Program, "Safety-Related Research in NASA's Aeronautics Research Mission Directorate: Overview," presentation to the committee, September 3, 2009.

the program leaders stated that they consult a spectrum of stakeholders from the aviation community, including the FAA, the JPDO, the National Transportation Safety Board (NTSB), CAST, other federal government agencies, and the aviation industry. The program office also reviews the recommendations in the *Decadal Survey of Civil Aeronautics,* other independent assessments of research needs by the NRC and others, and advice from the NASA Advisory Council.

The process of prioritization appears to be one in which important safety concerns in need of research are identified through the various external inputs described above and then assessed in light of NASA's existing safety research projects and capabilities (assets, workforce, and funding). Priority is given to those concerns that align well with the agency's existing research projects and capabilities. The objectives of these research projects and the utilization of existing research capabilities are then adjusted to better align with these emphasized safety concerns.

Based on this process, the Aviation Safety Program leadership stated that its safety research projects align with the following six overarching safety research concerns (listed in no particular order):

- New Operations,
- Flight In or Around Hazardous Conditions,
- Loss of Control,
- Durable Aircraft Structures and Systems,
- On-Board System Failures and Faults, and
- Analyzing Complex Systems for Safety.

NASA's safety research projects that align with each of these concerns, as well as their objectives, content, and resources, are examined in Chapter 3. What was not made clear to the committee, however, is if NASA undertakes objective evaluations of safety research priorities irrespective of its current research programs and capabilities. In general, NASA's safety research prioritization process appears to be heavily influenced by the availability of existing expertise and resources within the agency. Not having been provided with a more objective needs assessment, the committee is hindered in its ability to judge whether the six concerns are indeed the most appropriate priorities for federal investment going forward. The committee nevertheless decided to review each of these research concerns (as well as the safety research in the Fundamental Aeronautics Program and the Airspace Systems Program) with respect to each of the main aspects of the study charge, despite the fact that the committee could not weigh in on their overall appropriateness.

3

Review of Safety Research Objectives

Based on the process of research prioritization described in Chapter 2, NASA's aviation safety-related research programs are intended to align with the following six overarching, NASA-identified safety research concerns:

- New Operations,
- Flight In or Around Hazardous Conditions,
- Loss of Control,
- Durable Aircraft Structures and Systems,
- On-Board System Failures and Faults, and
- Analyzing Complex Systems for Safety.

In fiscal year (FY) 2010, these six research concerns received varying levels of Aviation Safety Program funding, as shown in Figure 3.1. Approximately half (49 percent) of the $60.1 million program budget goes to research addressing two concerns: (1) On-Board System Failures and Faults and (2) Loss of Control. The Integrated Vehicle Health Management (IVHM) and Integrated Resilient Aircraft Control (IRAC) projects contain most of this work. The IVHM project also houses the research conducted for Analyzing Complex Systems for Safety, which was 6 percent of the FY 2010 budget. The other three safety-related concerns, mostly covered by the Aircraft Aging and Durability (AAD) and Integrated Intelligent Flight Deck (IIFD) projects, receive 39 percent of program funds, while the remainder (6 percent) goes to program management. Budgeted totals for FY 2010 to FY 2014 (Table 3.1) show little change in the allocations across research projects.

Commensurate with the committee's statement of task, the research covering each of the six research concerns is examined with regard to (1) defining and prioritizing the research objectives, (2) resource allocation, (3) coordinating with the Federal Aviation Administration (FAA) and other relevant federal agencies and private entities, and (4) transitioning the research results from the programs into operational technologies and procedures and certification activities in a timely manner. The committee also attempted to review the safety-related research in the Fundamental Aeronautics Program and the Airspace Systems Program with regard to these four elements, although the documents and presentations reviewed by the committee lacked sufficient detail to enable as thorough an assessment. To answer these questions, the committee received numerous briefings from NASA and consulted the technical plans of the Aviation Safety Program's four main research projects, each of which addresses one or two of the six research concerns.

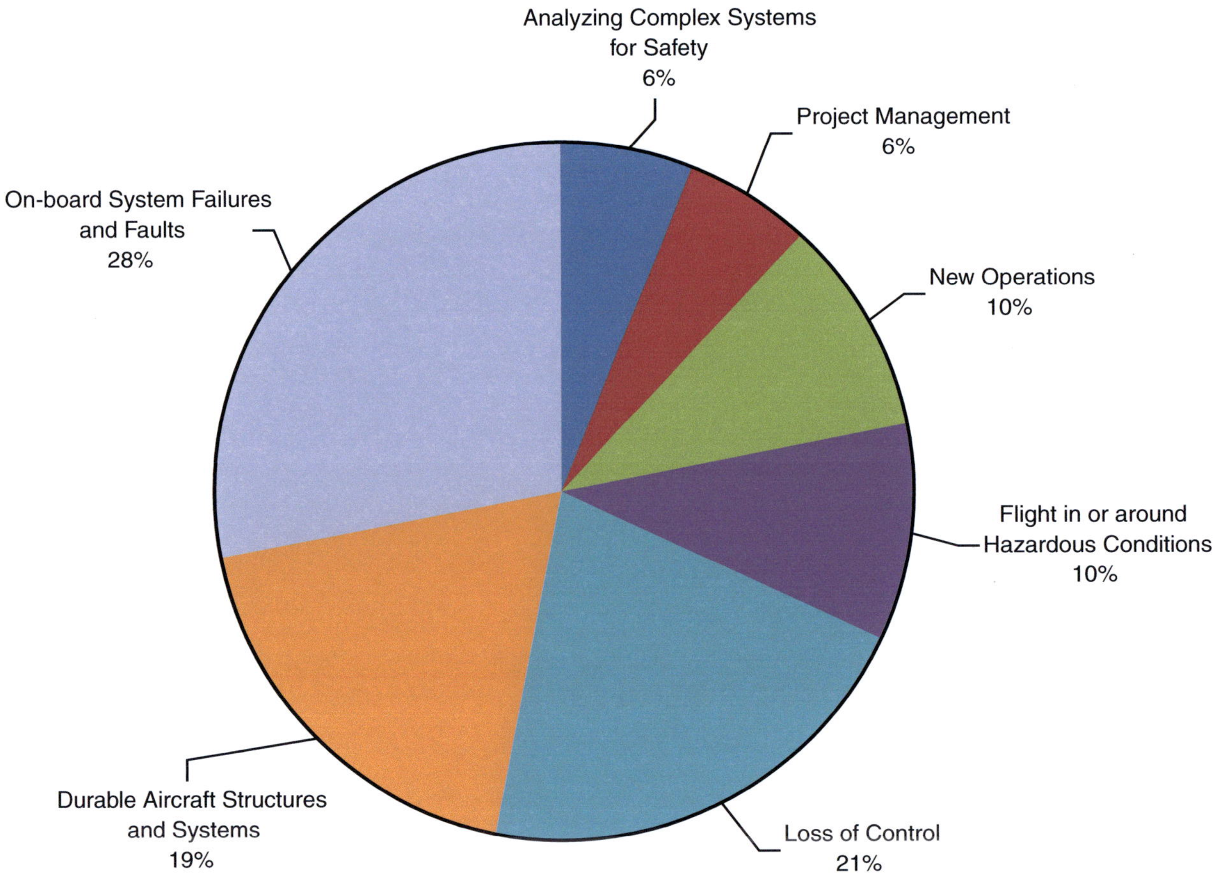

FIGURE 3.1 Fiscal year 2010 budget allocation by safety research concerns, Aviation Safety Program. SOURCE: Amy Pritchett, Director, NASA Aviation Safety Program, "Aviation Safety Program: Technical Overview," presentation to the committee, June 23, 2009.

TABLE 3.1 Aviation Safety Program 5-Year Budget, FY 2010 to FY 2014

	Millions of dollars				
	FY 2010	FY 2011	FY 2012	FY 2013	FY 2014
Aviation Safety Program Totals	60.1	59.6	59.2	61.7	62.5
Aircraft Aging and Durability	11.4	11.2	11.7	12.1	12.1
Integrated Intelligent Flight Deck	12.5	13.3	11.6	12.6	13.3
Integrated Resilient Aircraft Control	16.4	17.0	17.6	18.2	18.2
Integrated Vehicle Health Management	19.8	18.2	18.3	18.9	18.9

SOURCE: Amy Pritchett, Director, NASA Aviation Safety Program, "Safety-Related Research in NASA's Aeronautics Research Mission Directorate: Introduction to ARMD's Organization and Programmatic Overview," presentation to the committee, June 23, 2009.

NEW OPERATIONS

NASA cites the following safety challenges from the 2007 *National Plan for Aeronautics Research and Development and Related Infrastructure* (referred to as the "National Plan"),[1] as providing support for its research on New Operations:

- Rapidly and safely incorporating technological advances in avionics into aircraft;
- Understanding the key parameters of human performance in aviation to support the human contribution to safety; and,
- Ensuring safe operations for the complex mix of vehicles anticipated in the airspace system enabled by NextGen.

In addition, the agency cited 13 research and technology (R&T) safety challenges from the *Decadal Survey of Civil Aeronautics*[2] as being relevant to NASA's objectives in new operations research (see Appendix E, Table E.1). As a more general rationale for its work in this research concern, the agency pointed to its longstanding expertise in the study of future complex human-machine systems and operational concepts. NASA also believes it is particularly well suited for pursuing safety solutions in a more integrated manner, considering the roles and effects of new technology, procedures, human performance, and training.

Defining and Prioritizing Research Objectives

Most of the Aviation Safety Program's research on New Operations is mapped to the work in the IIFD project. This research is intended to support

- Robust, collaborative work environments;
- Effective robust automation-human systems; and
- Information management and portrayal for effective decision making.

NASA explained that these three objectives fit well within the Aviation Safety Program's mission to identify and design out the underlying causes of poor human and technology performance. To address these objectives, the IIFD research plan outlines four levels of research that will lead, at the highest level, to the development of flight deck systems that can improve safety. NASA argues that this will require foundational research on operator characterization and modeling, sensing, signal processing and hazard characterization, multimodal interfaces (which provide multiple modes of usage), and information interaction modeling, feeding into discipline-specific research on operator performance, enabling avionics, and design tools. Moving up another level, this research feeds into an understanding of robust human-automation systems and display and decision support.

The committee believes these research objectives are relevant to New Operations safety and that the individual research activities are generally responsive to the objectives. However, whether the research is sufficiently broad and how the research was programmed and prioritized within the objectives is less clear. For instance, one might have expected to find more work on novel vehicle configurations, such as Unmanned Aircraft Systems (UAS), the design of which may have safety implications.[3] With respect to the objective of supporting robust collaborative work environments, the committee was briefed on research being undertaken to develop a model-based design and evaluation tool for multiagent situation awareness. This particular work appears to be well suited to this objective; however, when looking across the IIFD project, there is little evidence of much other comparable work.

---

[1] National Science and Technology Council, *National Plan for Aeronautics Research and Development and Related Infrastructure,* Office of Science and Technology Policy, Executive Office of the President, Washington, D.C., December 2007, available at http://www.aeronautics.nasa.gov/releases/aero_rd_plan_final_21_dec_2007.pdf.

[2] National Research Council, *Decadal Survey of Civil Aeronautics: Foundation for the Future,* The National Academies Press, Washington, D.C., 2006.

[3] Note that the president's fiscal year 2011 budget includes funding for safety research on Unmanned Aircraft Systems.

With respect to research in support of robust human-automation systems, the committee observes a great deal of relevant work under way. In the case of the third objective—to support information management and portrayal—much of the work centers on information portrayal and less on information management.[4] While this third objective is appropriate, it implies a broader scope of work than information portrayal. One might expect, for instance, to find more research in areas such as algorithm development for integrating and fusing data.

More generally, the committee observes that there is a great deal of emphasis on the flight decks and interactions between pilots and on-board systems. Relatively little work applies to the performance of and interactions among people in other areas, such as air traffic control, maintenance, and airline operation centers, and with respect to ground-based automation. It is a failing on NASA's part that the research does not support joint situational awareness for the different roles within the system. In addition, the research tends to be oriented toward existing rather than emerging flight-deck automation. One would expect that research on New Operations would also consider novel vehicle configurations and the implications of their design on safety. These gaps imply a need for more collaboration with other Aviation Safety Program projects, especially Loss of Control (in IRAC), as well as ARMD's Airspace Systems Program and Fundamental Aeronautics Program.

Resource Allocation

Each of the New Operations research objectives receives varying levels of funding, as shown in Figure 3.2. In FY 2010, research pertaining to the objective of information management and portrayal received nearly half (48 percent) of the funds allocated for research on New Operations. One-third of the funds (33 percent) went to research in support of effective robust human-automation systems, and the remainder (19 percent) went to research on robust, collaborative work environments.[5] Whether this is an appropriate division of resources is difficult to judge. The committee questions why more research is not being undertaken on information management and processing since nearly half of New Operations research funding goes to the information management and portrayal objective.

Coordination with the FAA and Others

NASA indicated that the relevant agencies for coordinating research in this area are the FAA, the Department of Defense (DOD), and the Joint Planning and Development Office (JPDO) (as well as consultation with the National Transportation Safety Board [NTSB]). It appears the research is coordinated well with the FAA and JPDO and that there are many positive interactions taking place among FAA and NASA human factors researchers. Indeed, FAA-sponsored work may be as important to the New Operations objectives as some of the work sponsored internally by NASA (for example, the FAA's single-pilot workload study); hence, such collaboration is critical.

The committee notes that formal mechanisms for collaboration exist across all of the cited organizations, and finds this to be a strength of the research program. However, inasmuch as the primary focus of the work is the flight deck, this may limit opportunities for DOD collaboration in New Operations areas such as UAS.

Research Transition

NASA pointed to its participation on a number of government and industry committees pertinent to New Operations research. The Flight Deck Research Working Group (FDR WG), a joint effort of NASA and industry, is an example of one such effort. The agency also cited its use of Space Act agreements to partner with private companies, such as Boeing, to conduct research and transition the results to practice. Among the cited criteria

[4] The committee notes, for instance, that in Table 4-6 of the Integrated Intelligent Flight Deck (IIFD) Technologies Technical Plan only 3 of 10 projects (forward-looking remote sensing, imaging processing, and methods for exploiting cross-modality info transfer) focus on processing of information, and this is one of the few IIFD tables where information processing would be expected. See NASA, *Integrated Intelligent Flight Deck Technologies Technical Plan Summary (FY2009 – FY2013)*, Aviation Safety Program, Aeronautics Research Mission Directorate, NASA, Washington, D.C., March 13, 2009, available at http://www.aeronautics.nasa.gov/nra_pdf/ iifd_tech_plan_2009.pdf, Table 4-6, p. 36.

[5] Amy Pritchett, Director, NASA Aviation Safety Program, "Research Objective 1: New Operations," presentation to the committee, September 3, 2009, p. 20.

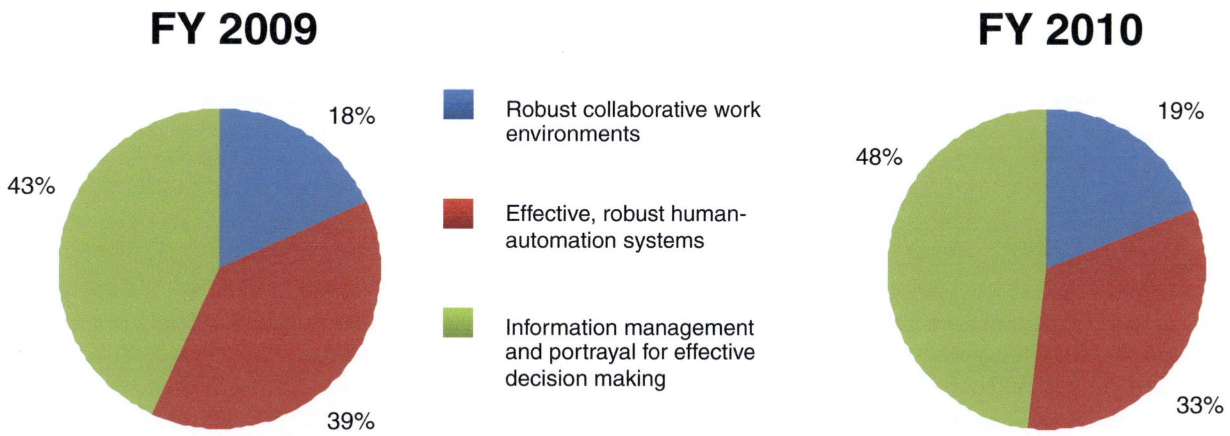

FIGURE 3.2 Research expenditures towards New Operations. SOURCE: Amy Pritchett, Director, NASA Aviation Safety Program, "Research Objective 1: New Operations," presentation to the committee, September 3, 2009.

for successful transitioning of the research results are their use as the basis for regulation and the extent to which researchers contribute to the technical literature. Indeed, the committee finds that NASA researchers have been prolific in publishing articles and books for the research and operational communities. Much of this work has focused on human factors and flight-deck interfaces, which are traditional areas of NASA expertise.

FLIGHT IN OR AROUND HAZARDOUS CONDITIONS

The focus of this research concern is avoidance of hazardous conditions through sensing and portrayal of environmental hazards, as well as specific research on aircraft icing, including modeling and sensing of airframe and engine icing. The topic is consistent with the National Plan's safety challenge to ensure safe operations for the complex mix of vehicles anticipated within NextGen. In briefings to the committee, NASA identified four safety challenges in the NRC *Decadal Survey* as relevant to the topic (see Appendix E, Table E.2). Commensurate with its mission to exploit its specialized expertise and assets, NASA noted that it has a well-respected workforce with significant experience in icing research (e.g., Glenn Research Center Icing Branch).

Defining and Prioritizing Research Objectives

NASA's research into this area is split across the IIFD and IRAC projects, with additional work being done in the Fundamental Aeronautics Program. Researchers are investigating new airborne sensing and alerting systems that extend the performance limits of weather radar technology and out-the-window human observations. This involves exploring forward-looking sensing methods, image processing, feature extraction, hazard characterization, and remote icing sensing, especially in conditions known as high ice water content (HIWC) conditions. Researchers are aiming for high-integrity detection of terminal area hazards, including objects on the runway, wake vortices, traffic, vertical obstructions, and terrain.

The committee determined that the specific research being conducted within this research concern appears to be appropriate, although the committee did not see any particular basis for NASA's prioritization of the research beyond a general inference based on its resource allocation. NASA and the committee recognize that improved airborne hazard detection and alerting continue to be a major safety need. Precise hazard detection and forecasting are expected to be important in NextGen, which should involve more deliberate and precise operations near hazardous conditions, thus requiring better sensing of and guidance near such conditions. Better sensing can also lead to capacity improvements; currently, a major capacity limitation in the system is wake turbulence-based spacing. Research into better wake sensing offers the possibility of safely improving airport capacity by reducing the time between takeoffs or landings on individual and closely spaced parallel runways.

NASA also recognizes the continuing hazards from in-flight icing for all aircraft classes and perceives a focused need for understanding fundamental issues in HIWC for turbine aircraft and a need for a fundamental examination of rotorcraft icing. NASA is particularly well situated to lead a national effort in this area through in-house expertise, funding strategic computational tool development, and partnering with industry for validation data. NASA has a historical strength in icing research; this area represents one of several key nodes of high-quality research in which NASA leads the nation and world.

HIWC is a recently identified hazard in deep convection and has been related to more than 100 power-loss events in turbine-powered aircraft. The meteorology and physical mechanisms that cause these events are weakly understood. NASA has been working closely with industry and other international research organizations to diagnose and understand HIWC hazards.

Rotorcraft are increasingly being used in all weather operations and have complex behavior in icing due to the variations in icing behavior between the rotating and non-rotating components as well as the potential for imbalance and extreme vibration dynamics of ice accretion and potential vibrations due to asymmetric ice accretion or shedding. While much of the prior industry research was focused on the topic of de-icing, there is now an emerging interest in the root causes and fundamental issues associated with ice formation and shedding.

From an organizational standpoint, it was not entirely clear to the committee why research into rotorcraft icing is separate from other icing research and a part of the Fundamental Aeronautics Program as opposed to the Aviation Safety Program. The committee did not see this organization as particularly problematic, but no specific rationale was given to explain it to the committee.

Resource Allocation

Figure 3.3 shows NASA's research expenditures toward Hazardous Conditions split between sensing and portraying environmental hazards and modeling and sensing airframe and engine icing and icing conditions. Altogether, research into Fight In or Around Hazardous Conditions represented around 9 percent of the Aviation Safety Program budget in both FY 2009 and FY 2010.[6]

Recognizing that NASA has chosen to focus on sensing and icing conditions, within this context the committee finds no basis to question this allocation of resources, including the FY 2010 increased emphasis on airframe and engine icing. However, the committee notes a concern regarding the availability of the Icing Research Tunnel at NASA Glenn Research Center. This facility, one of the largest refrigerated wind tunnels that duplicates natural icing conditions, is a critical resource for the government and industry but appears to be less available for use by NASA, which may impact the agency's research in this area.

Coordination with the FAA and Others

NASA briefed the committee on two key forms of research coordination—the HIWC team's partners and NASA's involvement with the Aircraft Icing Research Alliance (AIRA). The HIWC team partners with a variety of groups, including the FAA, whose focus is propulsion icing in HIWC environments; the Canadian National Research Council, to improve measurement capabilities in these conditions; and the Australian Bureau of Meteorology, to improve understanding of certain conditions that have led to engine power-loss events. The purpose of the AIRA is to coordinate collaborative aircraft icing research activities that could improve operations in icing conditions. The alliance is made up of many partners, including NASA, the FAA, the National Oceanic and Atmospheric Administration (NOAA), Environment Canada, Transport Canada, the Canadian National Research Council, and the U.K. Defense Science and Technology Laboratory.

Overall, the committee was particularly impressed by the integration and collaboration of research conducted

[6] Amy Pritchett, Director, NASA Aviation Safety Program, "Research Objective 2: Flight In or Around Hazardous Conditions," presentation to the committee, September 3, 2009, p. 16.

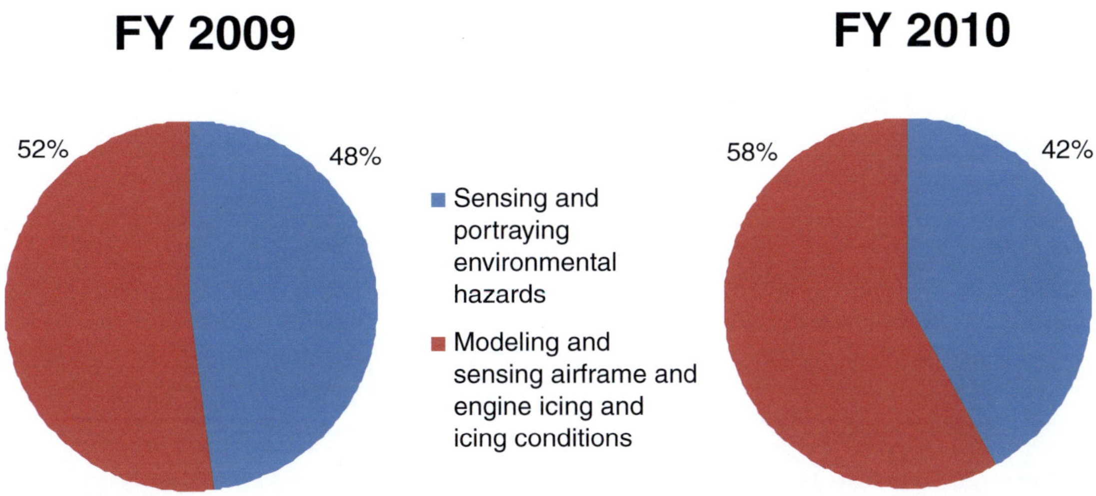

FIGURE 3.3 Research expenditures towards Flight In or Around Hazardous Conditions. SOURCE: Amy Pritchett, Director, NASA Aviation Safety Program, "Research Objective 2: Flight In or Around Hazardous Conditions," presentation to the committee, September 3, 2009.

on HIWC conditions. This level of collaboration could be viewed as a prime model of how NASA can keep its research relevant to the needs of other entities within the National Airspace System. NASA's involvement with AIRA, which also includes coordination of HIWC research, is also a strong mechanism to ensure that NASA's research in icing is properly coordinated with other relevant entities.

Research Transition

NASA pointed to several Space Act agreements with companies, such as Boeing and Goodrich, and the number of users of its LEWICE (Lewis ICE accretion program) software to demonstrate its ability to transition the results of its research in hazardous conditions. LEWICE codes are the industry standard for ice accretion simulation, and the software is used by hundreds in the aeronautics community. As with its research into New Operations, NASA's membership in several industry/government working groups, such as the FDR WG, enables partnerships that help align NASA's research in this area with the system users. Additionally, the committee was impressed with the large number of references and citations to NASA's icing research from researchers external to NASA.

LOSS OF CONTROL

As shown in Figure 3.4, loss of control is the most significant cause of fatal crashes by commercial jets worldwide and is a problem NASA believes may be exacerbated by future aircraft with different handling qualities and dynamics in upset conditions.

As further support for focusing on Loss of Control as a safety research concern, NASA pointed to several safety challenges in the National Plan that are relevant, including the following:

- Stabilizing and maneuvering next-generation aircraft in response to safety issues in the NextGen airspace,
- Rapidly and safely incorporating advances in avionics,
- Understanding and predicting system-wide safety concerns of the airspace system and the vehicles envisioned by NextGen, and
- Understanding the key parameters of human performance in aviation.

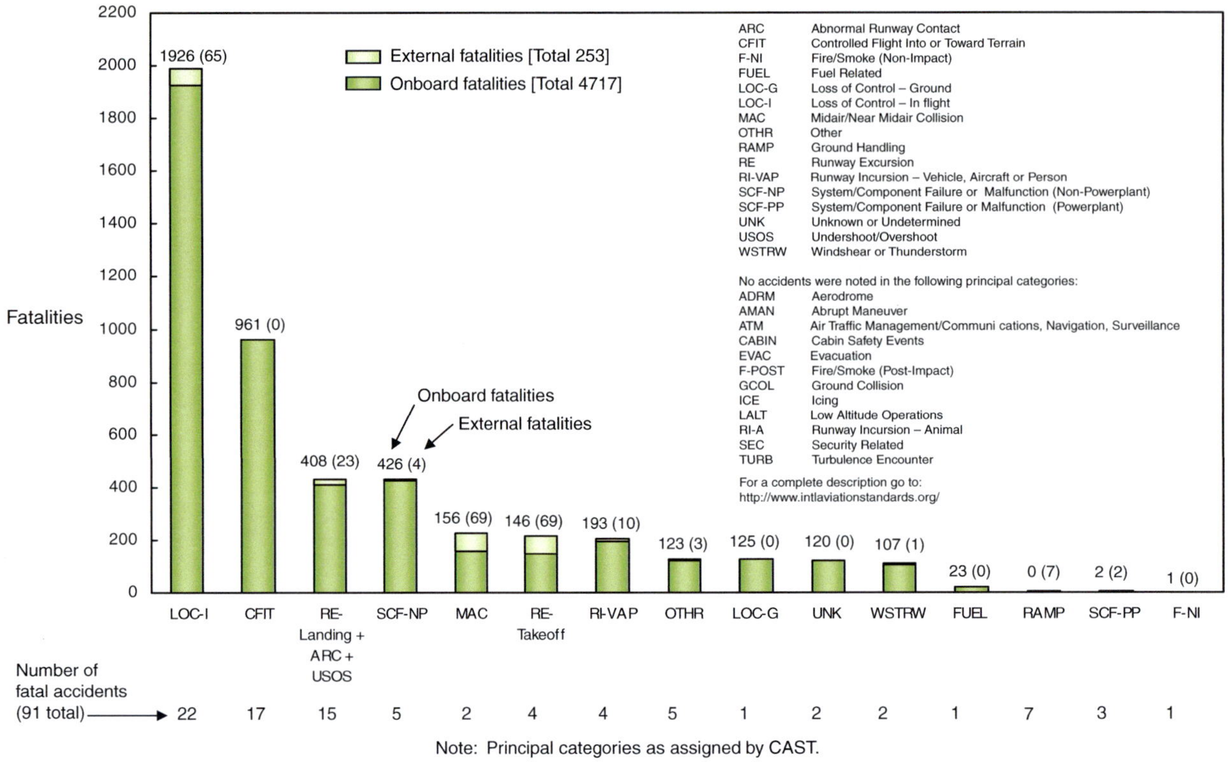

FIGURE 3.4 Fatal accidents by worldwide commercial jet fleet, 1999-2008, categorized by occurrence categories by the common taxonomy team of CAST/ICAO (CICTT). NOTE: The accident list is developed from the FAA, ICAO, and IATA statistics and reports. It covers worldwide Part 121 (and equivalent) jet operations (not turboprops). Each accident is assigned a single CICTT cause category. SOURCE: Boeing, Statistical Summary of Commercial Jet Airplane Accidents: Worldwide Operations, 1959-2008, Boeing Commercial Airplanes, http://www.boeing.com/news/techissues/pdf/statsum.pdf, July 2009.

NASA cited five relevant R&T challenges having high safety priority as identified in the NRC *Decadal Survey* (see Appendix E, Table E.3).

Defining and Prioritizing Research Objectives

NASA's research on Loss of Control is handled primarily in the Aviation Safety Program's IRAC project. According to IRAC's technical plan, this research seeks to arrive at a set of validated, multidisciplinary, and integrated aircraft control design tools and techniques that will advance the state of flight control to provide resilience for safe flight in the presence of adverse conditions.[7] This plan emphasizes that most of the challenges addressed are general in nature and thus applicable to a large class of aviation vehicles. The IRAC project is expected to advance understanding of the dynamics involved in loss of control, yielding an improved understanding, characterization, and prediction of conditions that threaten aircraft flight safety. The research is also intended to increase survivability and improve vehicle handling qualities under adverse conditions.

[7] See NASA, *Integrated Resilient Aircraft Control Technical Plan,* Aviation Safety Program, Aeronautics Research Mission Directorate, NASA, Washington, D.C., May 1, 2009, available at http://www.aeronautics.nasa.gov/nra_pdf/irac_tech_plan_c1.pdf.

The IRAC project is intended to be supportive of

- Avoidance of conditions conducive to loss of control,
- Detection of onset of loss of control, and
- Recovery from loss of control.

NASA explained that these three objectives fit well within the Aviation Safety Program's mission and that the agency has longstanding expertise in modeling flight dynamics, control theory, and flight validation of flight control mechanisms.

The committee concurs with both the need for NASA research on Loss of Control and the stated research objectives. The committee believes that the objectives themselves are well defined. Of concern, however, is that the types of loss of control cases being addressed by NASA do not appear to comprehensively address the class of problems being experienced in civil aviation. NASA does not provide a strong rationale for its interest in the loss-of-control problems being researched. In addition, the program of research gives limited attention to human factors and novel automation aids.

As indicated by Figure 3.4, loss of control is a major cause of aviation fatalities in the commercial sector. Many government and industry safety enhancement efforts have been introduced to address the problem, but with limited success. The committee concurs with NASA's assessment that this problem may be exacerbated in the future for various reasons, including the following:

- Airplane design for fuel efficiency that has led to less inherently stable aircraft (e.g., by shifting weight aft), which requires precise and timely control inputs by pilots or automation;
- Increasing aircraft automation to alleviate pilot workload and to minimize crew complement and reduce variability, which has reduced pilot situation awareness during all points in the flight; and
- Numerous observations (e.g., NTSB reports) that basic pilot skills have degraded over time, leading to an increased risk of loss of control during manually piloted flight.

To be sure, automation systems have improved aircraft-handling qualities and have produced safety benefits for commercial aviation, but automation has also introduced new safety challenges. The systems must be improved, both to reengage the pilot and to further provide assistance in high-workload situations that risk loss of control. The committee observes that in the IRAC work a great deal of emphasis is being placed on automation by advancing the state of the art of adaptive control as a design option to provide enhanced stability and maneuverability margins. *Adaptive control* refers to flight systems that can adapt and respond to rapidly changing conditions. Adaptive control as it functions in the presence of damage has been a popular topic of research in flight control design for aircraft safety and is one of the R&T challenges identified in the NRC *Decadal Survey.* Yet, whereas adaptive control has demonstrated recovery from loss of control in simulation and limited flight testing, it has not and will not be accepted until its behavior is better understood by pilots and deemed certifiable by the FAA for piloted aircraft.

The committee found an excessive focus on adaptive control in NASA's program and observed that the program does not recognize the importance of the interactions between the pilot and the automation as a significant cause of loss-of-control accidents. The adaptive control research does not adequately address the human element of control, nor does the adaptive control agenda foster the identification and development of alternative automation technologies that more broadly address loss-of-control challenges. As a general matter, human factors considerations do not appear to be sufficiently included in the Loss of Control research concern. NASA has traditionally had an important role in this area through its longstanding expertise in human factors and advanced flight control. This expertise, however, appears to have diminished.

As noted in the discussion earlier, the committee is also unclear about the level of collaboration among researchers addressing Loss of Control and New Operations safety. Both safety concerns entail automation and human factors challenges, suggesting the importance of research collaboration.

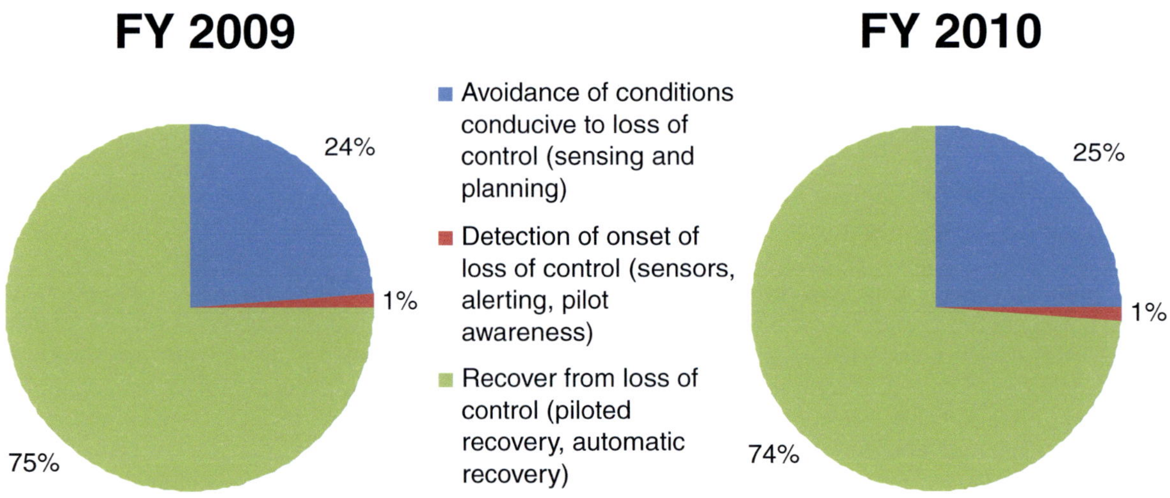

FIGURE 3.5 Research expenditures towards Loss of Control. SOURCE: Amy Pritchett, Director, NASA Aviation Safety Program, "Research Objective 3: Loss-of-Control," presentation to the committee, September 3, 2009.

Resource Allocation

Research for Loss of Control accounted for around 24 percent of the Aviation Safety Program budget in FY 2009 and 10 percent in FY 2010.[8] The three Loss of Control research objectives receive varying levels of funding, as shown in Figure 3.5. The distribution of resources is heavily weighted to the third objective in support of recovery from loss of control. In FY 2010, about 75 percent of the resources went to this objective, indicative of the project's focus on adaptive control. Giving greater attention to the human element and automation technology development beyond adaptive control would likely result in more resources being devoted to the objectives of avoiding conditions conducive to loss of control, detection of onset, and long-term assistance to the pilot in high-workload situations. The expense associated with flight testing adaptive control in full-scale flight test aircraft is substantial and may be taking resources away from these other objectives and from other plausible solutions to the loss-of-control problem.

Coordination with the FAA and Others

In briefings to the committee, NASA explained how it has coordinated its Loss of Control research with the FAA (through the Software and Digital Systems Safety Research Program), DOD, and industry (through the Joint Aircraft Survivability Program), as well as a number of North Atlantic Treaty Organization projects. Discussions with NASA presenters indicated to the committee that NASA is reevaluating and seeking to better coordinate its Loss of Control research efforts by consulting with other government agencies and industry. In particular, NASA has joined the CAST-led safety enhancement study associated with spatial disorientation and energy-state awareness. In the committee's view, such external coordination, along with collaboration with researchers on New Operations, is crucial to ensuring that research in this important area is programmed and executed well. In general, the committee observes that NASA's Loss of Control efforts associated with IRAC are being sufficiently coordinated with the FAA and others and its nascent efforts to consider other aspects of Loss of Control are promising.

[8] Amy Pritchett, Director, NASA Aviation Safety Program, "Research Objective 3: Loss-of-Control," presentation to the committee, September 3, 2009, p. 13.

Research Transition

In briefings to the committee, NASA expressed significant concerns about transitioning adaptive control results into commercial airplanes. Mechanisms do not exist for transitioning adaptive control into use, which presents major FAA certification challenges. NASA explained that it is still trying to understand the issues associated with safety certification. In the committee's view, an issue more fundamental than certification of adaptive control systems is the certification of non-deterministic software in general, a feature of several of the R&T challenges from the *Decadal Survey* (D14 and E1).[9] Nondeterministic software is not limited to single, prescribed outputs and may have great potential use in flight control systems, but because the results of its use can vary, the FAA has yet to develop certification standards for this type of software. Understanding the issues associated with this broader, overarching challenge, and seeking solutions, is an area to which NASA may be able to contribute through its safety research programs. While certification challenges represent a significant obstacle to the use of adaptive control systems per se, NASA's contribution to addressing these challenges can have significant benefits irrespective of whether these control systems are furthered.

DURABLE AIRCRAFT STRUCTURES AND SYSTEMS

Most of NASA's research on Durable Aircraft Structures and Systems is undertaken in the Aviation Safety Program's AAD research project. The AAD technical plan states that the aging properties of aircraft must be understood in order to design durable systems and support effective inspection and maintenance. By tapping its significant experience in materials science, damage mechanics, and nondestructive inspection, NASA sees its role in terms of building the fundamental underlying knowledge in this area, with a longer-term, more theoretical basis than the more applied and near-term work of the FAA, the DOD, and industry. In briefings to the committee, NASA pointed to three materials and structures R&T challenges in the NRC *Decadal Survey* that were rated high for safety (see Appendix E, Table E.4), as well as several other materials and structures R&T challenges that were deemed important for efficiency, capacity, or security, as motivation for the AAD project.

Defining and Prioritizing Research Objectives

The two main objectives of the AAD research program are (1) to gain full fundamental knowledge of existing, legacy aircraft and (2) to start on gaining knowledge about likely emerging materials and structures. To meet these objectives, AAD research activities are organized into three themes: Detect, Predict, and Mitigate. Research activities in each of these theme areas are listed in Box 3.1. Furthermore, the AAD project includes the following eight "challenge problems":

- Damage methodology for metallic airframe structures,
- Structural integrity of integral metallic structures,
- Durability and structural integrity of composite skin-stringer fuselage structure,
- Durable bonded joints,
- Durable engine fan containment structure,
- Durability of engine superalloy disks,
- Durability of engine hot section, and
- Wiring degradation and faults.

The committee believes challenge problems are an effective means of organizing and focusing the research, seemingly more useful than the three theme areas of Detect, Predict, and Mitigate. However, the research content of the AAD project is never presented in terms of the challenge problems; the research content and milestones are only presented in terms of the themes or in terms of a four-level approach to technology development (Foundational

[9] See National Research Council, *Decadal Survey of Civil Aeronautics: Foundation for the Future,* The National Academies Press, Washington, D.C., 2006, pp. 150-154.

BOX 3.1
Research Activities in Aircraft Aging and Durability Themes

Detect Theme

- Develop quantitative nondestructive evaluation (NDE) technologies that can measure the amount of fatigue in aerospace metals
- Develop eddy current NDE techniques for detection of deeply buried flaws in conventional metallic airframe components
- Collect electrical signatures data from chaffing of wiring laboratory interrogated wiring and also from using commercial microwave simulations
- Wire fault algorithm development
- NDE technique for interrogating butt-splice type crimped electrical connectors
- Sensing wire insulation effluence with gas chromatography
- NDE for cracks and residual stresses in integral metallic structures (high-speed machining, friction stir weld, electron beam free form fabrication)
- NDE techniques for flaw detection/characterization in composite structures
- Flaw detection and large area impact characterization for advanced composite fan cases
- Nanostructured materials for thin film high temperature sensors

Predict Theme

- Develop testing techniques to assess crack tip damage processes
- Assess intergranular and intragranular fracture in metals
- Produce robust closure models
- Develop atomistic models for inter and intra granular fracture of metallic materials
- Develop constitutive models for environmental damage in metals
- Three-dimensional finite element method (3D FEM) of non-linear cracks in integral structures
- Residual strength analysis in integral structures
- In-situ residual stress evaluation
- Develop 3D FEM models for damage in fiber metal laminates
- Fatigue characterization tests, delamination onset and growth, and life prediction tools for rotorcraft subcomponents
- Progressive damage methods for composite structures
- Structural integrity analysis of built-up composite structures
- Constitutive models and micro-macro mechanics models for strain-rate dependent constituent (resin) material deformation and failure analysis
- Finite element methods for composite structure impact mechanics analysis
- Micro-structural modeling of superalloy disk material

Mitigate Theme

- Aging of composite laminates
- Aging of bonded composites
- Materials aging: accelerated testing
- Development and characterization of improved adhesives
- Molecular modeling of epoxy-based materials
- Test technique for triaxial braid composites
- Composite material tests in aged condition
- Impact tests of aged composites for fan cases
- Superalloy disk rim durability assessment and enhancement
- Baseline hot corrosion characterization
- Develop ductile coatings for superalloy engine disks

SOURCE: Amy Pritchett, Director, NASA Aviation Safety Program, "Research Objective 4: Durable Aircraft Structures and Systems," presentation to the committee, September 3, 2009.

Research, Discipline-based Technology, Multi-Disciplinary Technology, and System-Level Application). As the AAD technical plan is currently written, it is impossible to separate the technology development path for one challenge problem from the path for another. Thus, the committee was unable to ascertain any prioritization of the challenge problems in the AAD project. In the event of a budget cut, it appears that the cut would be spread evenly among all the challenge problems. In other words, all of the challenge problems would be partially funded in the event of a budget cut, rather than fully funding high-priority problems and stopping work on the others.

How these particular challenge problems were developed is also unclear. It seems one of the primary criteria for selecting these challenge problems was to utilize the existing technical expertise and facilities at NASA. In the AAD technical plan, for instance, it is stated that the challenge problems were developed by NASA experts in the aeronautics and space shuttle programs, along with experts from the FAA and the DOD.[10] These experts "drew upon their experiences and vision for the future, to identify Challenge Problems that met the following criteria: aero-centric, safety critical, aging related, and containing technology development needs aligned with NASA's core capabilities and charter."[11]

The demonstrated need and contribution for each challenge problem is presented in the technical plan (see Box 3.2). The contributions are heavily weighted toward developing improved design methods, new structural designs, and less conservative material allowables to reduce the structural weight of aircraft, thus enabling more efficient and competitive use of aircraft. One constraint on these new design methods is to maintain current safety levels; this means that researchers might not be adventuresome in the directions that their research could take in order to ensure that no diminution of existing safety levels could result. While it is important to not diminish current safety levels, the committee observes that researchers in this area could approach safety performance as an opportunity—something that can be improved upon—rather than as a restriction.

In general, the committee observes that the AAD project is not as well directed toward addressing safety needs as might be expected. For instance, the committee did not observe significant work on the impact of maintenance and repair procedures on safety. Whereas durability is not normally considered a safety issue in and of itself, durability coupled with inappropriate maintenance is certainly a safety issue. In light of the fact that five of the eight challenge problems address durability, giving more consideration to maintenance and repair issues would seem warranted.

The committee also expected to find a strong connection between the AAD work and NASA work in integrated vehicle health management in light of the NRC *Decadal Survey* R&T challenges motivating the AAD project (see Appendix E, Table E.4). While this connection may exist, it was not evident from either the AAD or IVHM technical plans or briefings. In fact, the AAD technical plan seems to distance itself from integrated vehicle health management activities:

> The IVHM and AAD projects have very similar objectives, yet different approaches. AAD technology will enable more durable components and in many cases will rely on periodic inspections. With an IVHM system, inspections will be nearly continuous. Detection/diagnosis technologies developed in IVHM focus on light-weight, ubiquitous sensing such as fiber optics and MEMS, and on near-real-time interpretation of the extracted data, while the analogous technologies developed in AAD are ground-based and include full-field thermography and full-field ultrasonic imaging, processes that are impractical during flight. Similarly, the prognosis technologies in IVHM consider computational speed to be as important as accuracy, whereas AAD technologies are focused on elucidating the details of damage processes and achieving the greatest accuracy possible regardless of computational expense. Finally, mitigation technologies in IVHM are focused on in-situ and real-time mitigation of damage, while the technologies in AAD are oriented toward improving design and maintainability.[12]

It appears that the potential benefits of collaboration between the AAD and IVHM projects are not being realized at this time.

---

[10] NASA, *Aviation Safety Program Aircraft Aging and Durability Project Technical Plan Summary,* Washington, D.C., available at http://www.aeronautics.nasa.gov/nra_pdf/aad_technical_plan_c1.pdf, p. 2.

[11] NASA, *Aviation Safety Program Aircraft Aging and Durability Project Technical Plan Summary,* Washington, D.C., available at http://www.aeronautics.nasa.gov/nra_pdf/aad_technical_plan_c1.pdf, p. 2.

[12] NASA, *Aviation Safety Program Aircraft Aging and Durability Project Technical Plan Summary,* Washington, D.C., available at http://www.aeronautics.nasa.gov/nra_pdf/aad_technical_plan_c1.pdf, p. 9.

BOX 3.2
Aircraft Aging and Durability Challenge—Problems, Needs, and Contributions

Damage Methodology for Metallic Airframe Structures

Demonstrated Need: Aging involves complex interaction of load and environmental effects. Current environmental cracking solutions are empirical and of limited use for new material systems.

Contribution: Physics-based damage models for development of material design tools.

Structural Integrity of Integral Metallic Structure

Demonstrated Need: Increased use of integral metallic structures to reduce weight and part count. New selectively reinforced concepts competitive with composite structures.

Contribution: Predictive tools provide improved lifing/inspection protocols. Enable alternatives to composite structures.

Durability and Structural Integrity of Composite Skin-Stringer Fuselage Structure

Demonstrated Need: Increased composites and metal/composite hybrids in fuselage structure. Empirically based building-block design and certification approaches are expensive/highly conservative. Design margins not well quantified. Research pull by manufacturers, operators, NASA space programs, the Federal Aviation Administration.

Contribution: Reliable strength/life prediction methods accounting for aging degradation. Reduced empiricism in design and certification/reduced design cycle time.

Durable Bonded Joints

Demonstrated Need: Increased use of bonded joints to reduce part count. FAA: contamination/surface preparation is #1 problem in bonded structure. Recent examples of weak or degraded bonds causing structural failure. NTSB: Airbus rudder delamination associated with fluid ingression.

Contribution: Novel non-destructive inspection technology for bond strength. Chemistry approaches to improved adhesives/surface treatments. Mechanics predictions for degrading material.

Durability of Engine Fan Containment Structure

Demonstrated Need: New lightweight advanced composite concepts are under development (user pull). Fan containment is safety critical.

Contribution: Models of aging and configuration effects for optimal safety and efficiency. Improved "standardized" tools for blade-out event simulation and new non-destructive evaluation techniques enable acceptance of advanced concepts.

Durability of Engine Superalloy Disks

Demonstrated Need: Disk failures are rare but cannot be contained and are generally catastrophic. Before new disk alloys can be used at higher operating temperatures to improve operating efficiencies, their long-term durability must be understood and improved if necessary.

Contribution: Assured durability of new disk alloys at higher operating temperatures. Enables improved engine efficiency while retaining safety.

Durability of Engine Hot Section

Demonstrated Need: Very difficult to model turbine blade temperatures, strains, heat fluxes; measurements are needed. The turbine section has been consistently responsible for >$40 million per year in losses to the Air Force (second leading cause of aircraft damage).

Contribution: New bulk and nano-structured sensor materials and novel thin film harsh environment sensors provide high temperature characterization for use in component durability improvement.

Wiring Degradation and Faults

Demonstrated Need: Electrical failures from shorts, opens, insulation degradation, or overloaded circuits can have catastrophic effects in legacy and new vehicles. Attempts to contain damage after system failure don't always work.

Contribution: Predictive tool for condition-based maintenance reduces risk. Next-generation sensored/fault-tolerant distributed wiring designs.

SOURCE: NASA, *Aviation Safety Program Aircraft Aging and Durability Project Technical Plan Summary,* Washington, D.C., available at http://www.aeronautics.nasa.gov/nra_pdf/aad_technical_plan_c1.pdf.

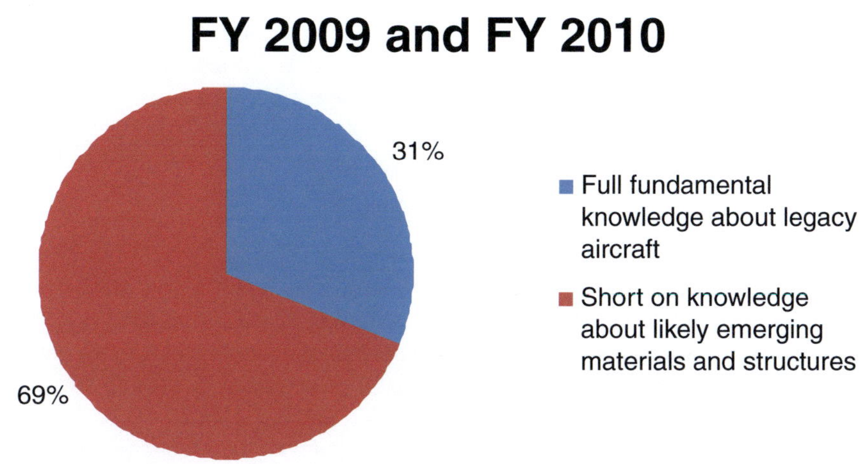

FIGURE 3.6 Research expenditures towards Durable Aircraft Structures and Systems. SOURCE: Amy Pritchett, Director, NASA Aviation Safety Program, "Research Objective 4: Durable Aircraft Structures and Systems," presentation to the committee, September 3, 2009.

Resource Allocation

Figure 3.6 shows NASA's research expenditures on Durable Aircraft Structures and Systems split between fundamental research about legacy aircraft and research into emerging materials and structures (the percentages are the same for FY 2009 and FY 2010). Altogether, research in this area accounted for 17 percent of the Aviation Safety Program budget in FY 2009 and 19 percent in FY 2010.[13]

About two-thirds of the R&D funds devoted to the AAD project are used for gaining knowledge about likely emerging materials and structures, mostly composites, and the remainder for gaining full, fundamental knowledge about legacy aircraft. Given NASA's longer-term research role, this allocation across objectives seems to be appropriate, as do the allocations by challenge problem.

The distribution of R&D funds across the theme areas has slightly more than half the funding directed toward the Predict theme, a quarter of the budget in the Detect theme, and the remainder in the Mitigate theme. It is not clear whether this distribution of funding across the themes resulted from individual researchers' interests or a conscious allocation process. However, given the previous observations regarding the importance of maintenance and repair procedures for safety, the committee expected to see slightly greater emphasis on the Mitigate theme.

More than three-fourths of AAD personnel are engaged in the four challenge problems dealing with airframe applications. Almost one-fifth of personnel are involved with the three challenge problems dealing with propulsion applications. The remaining 6 percent of personnel are working on the last challenge problem on wiring. Once again, it is not clear whether this distribution of the workforce was through self-selection of research activities or allocation by management. This allocation of personnel also appears consistent with the Aviation Safety Program's emphasis on large commercial aircraft.

Coordination with the FAA and Others

The research conducted in the AAD project appears to be coordinated with that of the FAA and other relevant federal agencies, mainly through working groups. As a result of a long history of NASA work in airframe structures and engines, AAD researchers are members of a number of industry and government working groups, including

[13] Amy Pritchett, Director, NASA Aviation Safety Program, "Research Objective 4: Durable Aircraft Structures and Systems," presentation to the committee, September 3, 2009, p. 15.

the Joint Council on Aging Aircraft, the Propulsion Instrumentation Working Group, the Interagency S&T Wiring Group, and the Joint Air Logistics Command (JALC) Non-destructive Inspection (NDI) Working Group. AAD researchers are also aware of relevant research going on within the FAA and the U.S. Air Force (USAF). It is difficult to determine the intensity of interactions and the degree of coordination from the AAD technical plan and the presentations made to the committee. However, the committee recognizes that the AAD project is making an effort to coordinate its activities with other government agencies and industry.

Research Transition

The committee notes that NASA has achieved considerable success with transitioning technology developed under the Detect theme. A device and method for inspecting crimp connections have been patented, and the aviation industry has shown interest in obtaining a license. Industry is interested in demonstrating NDI methods and sensors in realistic applications.

Technologies developed under the Predict theme are design-oriented, and transition therefore occurs through information exchange, publications, and participation in technical organizations. The adoption of modeling and analysis technology by industry and certification agencies can require years of demonstration to train people in the methods and build confidence. NASA is doing everything possible to stay engaged in the long transition process: establishing agreements with industry to demonstrate the technology, participating in organizations that develop standards, and publicizing developments at conferences and in journals.

In the Mitigate theme, NASA has been developing improved fabrication processes, standards for material certification, and new structural concepts. NASA has successfully transitioned technology here as well, especially in the area of composite containment cases for turbine engines.

ON-BOARD SYSTEM FAILURES AND FAULTS

Research for On-Board System Failures and Faults centers on four key objectives:

- Detection of system anomalies and adverse events,
- Diagnosis of causal factors, assessing severity of and distinguishing adverse events,
- Prognosis of remaining useful life, and
- Mitigation of impact of adverse effects to continue safe flight and landing.

When asked to explain what prompted the Aviation Safety Program's interest in this research concern, NASA briefers referred to the National Plan's safety challenges of "monitoring and assessing aircraft health at both the material and component level"[14] and "ensuring safe operations for the complex mix of vehicles anticipated within the airspace system enabled by NextGen."[15] They also cited five relevant research challenges with high safety priority in the *Decadal Survey,* as listed in Appendix E, Table E.5. Going forward, the agency expressed concern that failures at the component, subsystem, and system levels will remain a significant source of risk in NextGen aircraft employing new materials, using more electric and power systems, operating for longer periods under harsh environments, and depending on more complex software systems.

[14] National Science and Technology Council, *National Plan for Aeronautics Research and Development and Related Infrastructure,* Office of Science and Technology Policy, Executive Office of the President, Washington, D.C., available at http://www.whitehouse.gov/administration/eop/ostp/library/archives, p. 23.

[15] National Science and Technology Council, *National Plan for Aeronautics Research and Development and Related Infrastructure,* Office of Science and Technology Policy, Executive Office of the President, Washington, D.C., available at http://www.whitehouse.gov/administration/eop/ostp/library/archives, p. 24.

Defining and Prioritizing Research Objectives

The committee determined that the research objectives in this area are appropriate and well defined at a high level. However, the objectives are very broadly interpreted at the level of specific research activities to the extent that the alignment of some activities, such as ice crystal sensing or physics-based models of sensing performance, is somewhat questionable. The ice crystal and physics-based model activities appear to be "attached" to the project rather than aligned to the objectives.

Interested in knowing more about the rationale for research within On-Board System Failures and Faults, the committee consulted the technical plan for IVHM. This program is aimed at "[developing] validated tools, technologies, and techniques for automated detection, diagnosis and prognosis that enable mitigation of adverse events during flight…, [including] those that arise from system, subsystem, or component faults or failures due to damage, degradation, or environmental hazards."[16] As support for this research emphasis, the plan points out that system and component failures and malfunctions contribute to 24 percent of on-board fatal accidents and are underlying factors in many of the 26 percent of fatal accidents caused by loss of control.[17] Yet, apart from providing such general statistics, the plan offers little information supporting the specific concerns over increasing risks from failures and malfunctions in NextGen. The plan does not, for instance, analyze incident data or literature from academia and industry to extract more detailed information about causal factors from current aircraft incidents that might yield indications of future problems. Marshaling such information at the outset of the program would have provided a more specific basis for defining and justifying the research objectives in the detection, diagnosis, prognosis, and mitigation of on-board system failures and faults. In addition, the stated objectives in the plan were inconsistent with the objectives presented in the corresponding briefings given by NASA to the committee.

Resource Allocation

Figure 3.7 shows NASA's research expenditures into On-Board System Failures and Faults split between integration, detection of system anomalies and events, diagnosis of causal factors, prognosis of remaining useful life, and mitigation of impact of adverse effects. Altogether, research in this area accounted for 25 percent of the Aviation Safety Program budget in FY 2009 and 28 percent in FY 2010.[18]

Whether this is an appropriate division of resources is difficult to judge, but in general, the committee viewed this division as likely balanced and reasonable. The committee observed an impressive level of integration across the centers and branches for research for this area.

Coordination with the FAA and Others

NASA indicated a variety of agencies with which it coordinates its research in this area, including the FAA, JPDO, and the U.S. Air Force Research Laboratory (AFRL). The committee observed that the cooperative activities with the AFRL appear quite solid and are aligned to the high-level goals of the research into On-Board System Failures and Faults. These agreements span several types of research, including vehicle-level architecture, ground-to-flight architectures, IVHM-enabled condition-based maintenance, and data mining. NASA's principle coordination with the FAA in this area centers on participation in several FAA-sponsored teams and committees. The committee observes that NASA's participation aligns with its historical areas of expertise, such as in lightning and high-intensity radio frequency. While this is commendable, the committee is concerned that NASA may be limiting its involvement to those traditional areas and not positioning itself to address future, emerging concerns. NASA's coordination activities with JPDO in this area were somewhat vague and were thus difficult to evaluate.

---

[16] NASA, *Integrated Vehicle Health Management Technical Plan, Version 2.03,* Aviation Safety Program, Aeronautics Research Mission Directorate, NASA, Washington, D.C., November 2, 2009, available at http://www.aeronautics.nasa.gov/nra_pdf/ivhm_tech_plan_c1.pdf, p. 3.

[17] NASA, *Integrated Vehicle Health Management Technical Plan, Version 2.03,* Aviation Safety Program, Aeronautics Research Mission Directorate, NASA, Washington, D.C., November 2, 2009, available at http://www.aeronautics.nasa.gov/nra_pdf/ivhm_tech_plan_c1.pdf, p. 6.

[18] Amy Pritchett, Director, NASA Aviation Safety Program, "Research Objective 5: On-Board System Failures and Faults," presentation to the committee, September 3, 2009, p. 1.

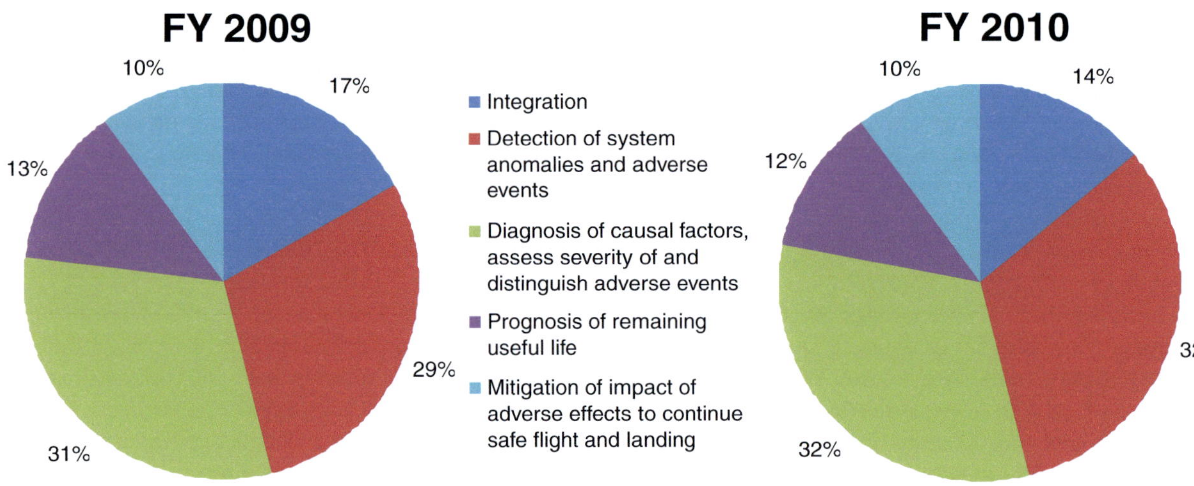

FIGURE 3.7 Research expenditures towards On-Board System Failures and Faults. SOURCE: Amy Pritchett, Director, NASA Aviation Safety Program, "Research Objective 5: On-Board System Failures and Faults," presentation to the committee, September 3, 2009.

Research Transition

NASA briefers did not give any evidence of any specific mechanisms or examples of technology transition in presentations to the committee. NASA did discuss several established partnerships with industry, such as with ExpressJet Airlines, a partnership that supports information sharing and continued use by NASA of engine performance data. However, the committee did not observe any significant transition activity within these partnerships. Another established partnership cited by briefers in their transition plans is with Boeing for validation of LEWICE software, but this partnership better fits with NASA's Hazardous Conditions research concern, so the merits of including it as part of the On-Board System Failures and Faults research concern were questionable to the committee.

ANALYZING COMPLEX SYSTEMS FOR SAFETY

NASA's interest in Analyzing Complex Systems for Safety encompasses (1) monitoring and predicting potential safety issues from operational data, (2) validating system requirements for safety objectives, and (3) verifying that designs meet system safety requirements. NASA explained to the committee that its interest in monitoring and predicting safety issues derives from a general recognition of the complexity of the airspace system for safety analysis and from the vast amounts of information available from current flight operations that can aid in monitoring for potential safety issues. The agency also noted that it has a charter to administer the Aviation Safety Reporting System and a continuing role in developing methods and tools for furthering safety data analysis capabilities. The agency's interest in verification and validation derives from concern that current safety assurance methods do not scale well to the complexity of current and potential future systems, resulting in costly analyses for innovative and novel designs such as those emerging from NextGen. The agency also noted that the JPDO has identified the gap in verification and validation (V&V) methods and tools for NextGen airspace and aircraft that NASA has accepted responsibility to fill. As an additional basis for pursuing these avenues of research, NASA cited 13 NRC *Decadal Survey* safety challenges (see Appendix E, Table E.6) and 5 of the National Plan's challenges.

Defining and Prioritizing Research Objectives

Recent research on monitoring and predicting potential safety issues is being undertaken in the IVHM project, where research is seeking to develop probabilistic models of potential fault and failure modes and data mining algorithms for analyzing large heterogeneous data sources from the current fleet. NASA has agreements with two airlines—EasyJet and Southwest Airlines—to test the data-mining methods under development for fleet-wide anomaly detection. With respect to this research effort, the committee did not find a strong rationale for it in the IVHM plan. The committee suspects that obtaining and analyzing operational data in an environment lacking a standardized data collection system among operators will present significant challenges to such analytic efforts. Greater recognition of this challenge in the IVHM technical plan is warranted to ensure appropriate research objectives and approaches. It was not clear to the committee how the cited safety challenges from the *Decadal Survey* or the National Plan informed the research objectives within this research concern, and overall, the process NASA used to identify objectives was obscure. In addition, the committee was not presented with a rationale for the specific airlines NASA has formed agreements with and suspects that these airlines may be poorly suited as testers of this technology, given their somewhat atypical complement of aircraft compared to larger airlines. The selection of these airlines seems rather ad hoc to the committee, and the committee observed that NASA does not appear to have a systematic approach to engage with airlines to obtain the operational data that would be necessary for effective safety analysis in this research area.

More recently, the agency, along with the JPDO, has recognized that validation of system requirements relative to safety objectives and the verification that designs meet system safety requirements necessitate a focused approach. The JPDO's integrated workplan calls for NASA to take the lead in this endeavor, and a focus area in V&V under the Aviation Safety Program of ARMD has been developed covering three research centers. The plan for executing the research objectives in this area was still under development when the committee met with NASA management and researchers. The committee observed that the plan is evolving in promising directions, although it notes some reservations. Research on fault detection, diagnosis, and classification is being performed with a high reliance on algorithms, and in doing so, the research is failing to utilize human capabilities in pattern recognition and the essential human role in data mining. Additionally, the objectives of the technical plan do not align with the objectives espoused in presentations made by researchers at the Aviation Safety Program's November 2009 technical conference, raising concerns of a serious disconnect between the high-level goals in the technical plan and the objectives of the individual research activities.

ARMD research activities on V&V remain in formulation and are expected to transcend the four main Aviation Safety Program research projects. V&V research is intended to meet the JPDO's needs in support of NextGen: to demonstrate advanced methods to address new safety analysis assessment challenges relevant to the broader aviation community, to reduce barriers to innovation associated with safety V&V, and to develop V&V methods for safety throughout the entire life cycle. The committee considers this an important avenue of research for NASA to be engaged in, one that is meeting a specific need of JPDO and is not being pursued elsewhere. It is distinctly possible that V&V could become another of NASA's key nodes of expertise and that NASA will lead the nation or world in high-quality research in this area.

Overall, the committee viewed the objectives of NASA's research in Analyzing Complex Systems for Safety as appropriate, but not well defined or prioritized. The committee suspects that these failings are due in large part to the ongoing development of a new technical plan and new research directions. The decision to begin work on V&V demonstrated to the committee that NASA's research portfolio can be responsive to the specific needs of the broader aviation safety community.

Resource Allocation

Figure 3.8 shows NASA's research expenditure toward Analyzing Complex Systems for Safety split into monitoring and predicting potential safety issues from operational data, validating system requirements, and verifying that designs meet safety requirements. Altogether, research in this area accounted for 7 percent of the

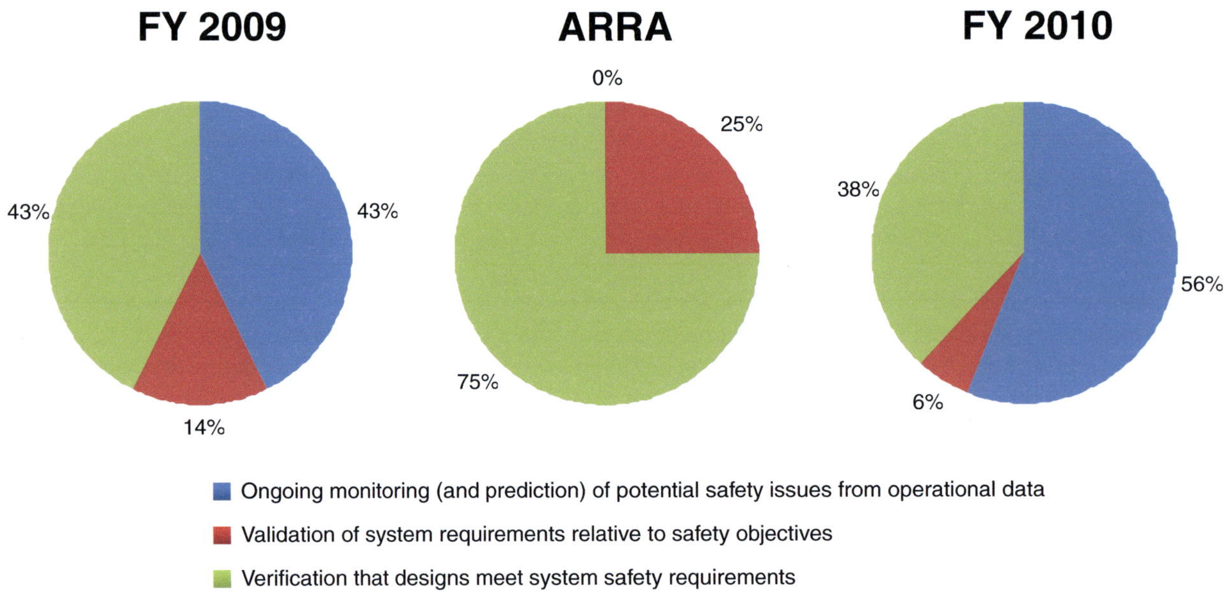

FIGURE 3.8 Research expenditures towards Analyzing Complex Systems for Safety. SOURCE: Amy Pritchett, Director, NASA Aviation Safety Program, "Research Objective 6: Analyzing Complex Systems for Safety," presentation to the committee, September 3, 2009.

Aviation Safety Program budget in FY 2009 and 6 percent in FY 2010.[19] Funds from the American Recovery and Reinvestment Act (ARRA) were also invested into this area.

The committee has some concerns regarding the allocation of funds for research in this area. The ARRA funds were designated for exclusive use for V&V, and a large expenditure on this new area is appropriate in the committee's view. However, this was not a lasting investment, and the overall funding for validation dipped rather significantly from 15 percent in 2009 to a mere 6 percent in 2010. Because of the highly variable funding, the committee was unsure of NASA's long-term commitment to V&V research, although this concern may be mitigated if the president's FY 2011 budget is adopted, since it commits $20 million per year for 5 years to V&V research.[20]

The committee also observed that NASA Research Announcement (NRA) and Small Business Innovation Research (SBIR) contracts are being used to fill large expertise gaps in V&V at NASA. In general, NRAs are used by NASA to supplement its research activities, and recipients of these contracts "largely manage their own research projects with minimal oversight by the agency."[21] NASA uses SBIR contracts for the development of "a technology in response to a specific set of NASA mission driven needs."[22] The committee did not see any evidence that NASA was actively trying to grow its V&V core competency, which would be necessary for NASA to take the leadership research role required of it by the JPDO.

[19] Amy Pritchett, Director, NASA Aviation Safety Program, "Research Objective 6: Analyzing Complex Systems for Safety," presentation to the committee, September 3, 2009, p. 9.

[20] Jaiwon Shin, Associate Administrator, NASA Aeronautics Research Mission Directorate, "NASA Aeronautics Research," presentation to the Aeronautics and Space Engineering Board, March 9, 2010, p. 7.

[21] NASA, *Guidebook for Proposers Responding to a NASA Research Announcement (NRA) or Cooperative Agreement Notice (CAN)*, Washington, D.C., January 2010, available at http://www.hq.nasa.gov/office/procurement/nraguidebook/proposer2010.pdf, p. 2.

[22] See NASA's Small Business Innovation Research and Small Business Technology Transfer Programs Web site, "First Time Participants," at http://sbir.gsfc.nasa.gov/SBIR/ftp_faq.html.

Coordination with the FAA and Others

The committee observed many different activities for coordination in this area with the FAA, the JPDO, and various companies and universities. NASA is partnering with the other members of ASIAS to identify data-mining needs. Through an agreement with the FAA, NASA is also still in charge of maintaining the Aviation Safety Reporting System. More recent collaborations in this area include NASA's attendance at various JPDO and other interagency meetings, the launching of the DASHlink Web site (a novel collaboration and communication tool for the FAA and industry, particularly for V&V), and NASA's continued use of NRA and SBIR contracts.[23] The committee viewed this level of coordination as strong, although there is still opportunity for improvement, such as acquisition of expertise in V&V of complex systems or joint NASA/contractor work that would build NASA core competence.

Research Transition

NASA briefers pointed to four agreements, three of which are currently in development, as examples of its mechanisms to transition research results in this area:

- *EasyJet:* Research, development, and deployment of data-mining algorithms to support fleet-wide operational anomaly detection;
- *ONERA:* Systematic verification and validation of data-mining algorithms on heterogeneous data sources (in development);
- *Sagem:* Evaluation of data-mining algorithms in multi-airline and rotorcraft data sources (in development); and
- *Southwest Airlines:* Research, development, and deployment of data-mining algorithms to support fleet-wide operational anomaly detection (in development).[24]

The committee had difficulty seeing these agreements as an actual transitioning of research results. NASA is not giving these companies specific technologies that they can use; the agreements seem more focused on simply allowing NASA to access the companies' data for the development and testing of NASA's data-mining algorithms. However, transition plans for the new V&V research were under development and were thus not specifically evaluated by the committee.

SAFETY-RELATED RESEARCH IN THE FUNDAMENTAL AERONAUTICS PROGRAM

Recognizing that its charge extends beyond the research being conducted in NASA's Aviation Safety Program, the committee heard briefings regarding safety-related research within the Fundamental Aeronautics Program. Unfortunately, the committee could not delve into the research as thoroughly as it was able to examine the research in the Aviation Safety Program; thus, its assessment is at a somewhat higher level than the six research concerns described above.

Defining and Prioritizing Research Objectives

The primary focus of NASA ARMD's Fundamental Aeronautics Program is on performance, efficiency, and environmental challenges for future air vehicles; safety is "generally not the focus of research."[25] In order to ensure that safety concerns are addressed, researchers coordinate with the Aviation Safety Program as needed. NASA

[23] See NASA's DASHlink Web site at https://c3.ndc.nasa.gov/dl/.

[24] Amy Pritchett, Director, NASA Aviation Safety Program, "Research Objective 6: Analyzing Complex Systems for Safety," presentation to the committee, September 3, 2009, p. 9.

[25] Jay Dryer, Director, Fundamental Aeronautics Program, "NASA Fundamental Aeronautics Program," presentation to the committee, November 19, 2009, p. 10.

briefers pointed to two areas of current safety-related research within the Fundamental Aeronautics Program: flight in or around hazardous conditions and rotorcraft crash survivability. The briefers cite some of the mobility goals of the National Plan as providing support for its research on rotorcraft in general. Additionally, briefers pointed to NRC *Decadal Survey* challenges (see Appendix E, Table E.7) as well as industry input and NASA system studies as further rationale for the research.

Safety researchers in the Fundamental Aeronautics Program focus on four specific subactivities:

- Development of new techniques in grid generation and computational fluid dynamics;
- Testing in small-scale wind turbine tunnels to provide critical validation data for specific areas, such as dynamic stall, active flow control, and airfoils with ice accretion;
- Development and demonstration of an externally deployable energy absorber concept for improved crash protection; and
- Validation of analytical models that focus on craft crashworthiness, including characterization of dynamic material properties, modeling of human occupants and predicting injury, multi-terrain impact simulation, development of fully integrated simulation models, and code validation studies that focus on probabilistic analysis and uncertainty quantification.

In most cases, the committee observed that these subactivities were well defined and appropriate safety-related research. The committee notes positively that much of the hazardous conditions research in the Fundamental Aeronautics Program is coordinated with researchers in the Aviation Safety Program, thus reducing overall inefficiencies.

The committee, however, had some concerns regarding crash survivability. NASA briefers pointed to *Decadal Survey* challenge C8, "structural innovations for high-speed rotorcraft," as one of the two primary inputs or rationales for the research. However, the *Decadal Survey* notes that this challenge "involves the use of many disruptive technologies and is unlikely to significantly increase safety or reliability."[26] The challenge received a "1" in the category of *safety and reliability*—the lowest score possible. Aircraft crash survivability is one of the goals in the National Plan's safety objectives, but this goal is not specific to rotorcraft; the committee thus questions NASA's focus. Overall, the committee was unconvinced of the specific value of this research for enhancing the safety of the NextGen system.

As with several of the research concerns discussed previously in this chapter, the committee was unable to determine how NASA prioritized its safety-related research within the Fundamental Aeronautics Program.

Resource Allocation

Figure 3.9 shows the aviation safety-related research expenditures in ARMD's Fundamental Aeronautics Program divided into the two primary areas of hazardous conditions and crash survivability.

Total expenditures equaled about $2.8 and $3.2 million, respectively, for FY 2009 and FY 2010. These represent approximately 0.9 percent and 1.4 percent of the total budget of the Fundamental Aeronautics Program for FY 2009 and FY 2010.[27] The low percentage was not entirely surprising to the committee, as the primary focus of the Fundamental Aeronautics Program is not on safety-related research. Based on the documents and presentations reviewed by the committee, it was not possible to determine whether the resource allocation toward safety-related research in the Fundamental Aeronautics Program was adequate or appropriate.

---

[26] See National Research Council, *Decadal Survey of Civil Aeronautics: Foundation for the Future,* The National Academies Press, Washington, D.C., 2006, p. 122.

[27] Based on numbers in Jay Dryer, Director, Fundamental Aeronautics Program, "NASA Fundamental Aeronautics Program," presentation to the committee, November 19, 2009, pp. 20 and 31.

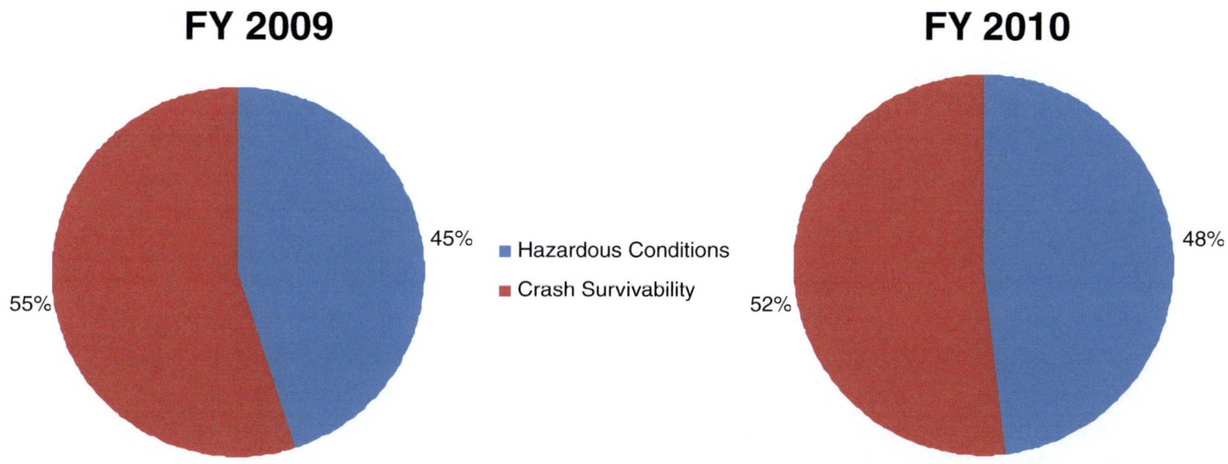

FIGURE 3.9 Research expenditures towards aviation safety-related research in the Fundamental Aeronautics Program. NOTE: The following caveats were included on the data in presentations to the committee: (1): Center for Rotorcraft Innovation funding in support of collaborative work is proprietary and is not shown; (2) NRA funding was completed with FY 2009 dollars. SOURCE: Based on numbers in Jay Dryer, Director, Fundamental Aeronautics Program, "NASA Fundamental Aeronautics Program," presentation to the committee, November 19, 2009, pp. 20 and 31.

Coordination with the FAA and Others

The Fundamental Aeronautics Program's safety-related research is coordinated with the FAA and others primarily through industry-government agreements, ranging from membership in working groups, such as the CMH-17 Crashworthiness Working Group, to participation in research centers, such as the National Rotorcraft Technology Center. The committee observed that these coordination activities are appropriate for the specific research areas, although the committee was unable to assess the depth of these activities. Based on the briefings given to the committee, safety researchers in the Fundamental Aeronautics Program do not appear to engage directly with the FAA in one-on-one collaboration outside of the working groups or research centers. The committee questions the safety premise behind the crashworthiness research, but observes that if the research is warranted (either for safety or for other reasons), NASA's coordination activities in the area are appropriate.

Research Transition

There was a distinct lack of detail in NASA's presentations to the committee regarding the research transition mechanisms and methods of the Fundamental Aeronautics Program. NASA simply asserted that either of these research activities will be a success if one is "established, validated and becomes widely used throughout industry within the next 10 years."[28] How NASA plans to get the technology to the industry was not presented to the committee; therefore, the committee cannot make a true assessment of NASA's transition plans in this area other than to observe that there appears to be a stark lack of specifics. Some examples of research transition in icing research were provided, but these examples were similar to and contained fewer details than the transition efforts described to the committee for icing research in the Aviation Safety Program.

[28] Jay Dryer, Director, Fundamental Aeronautics Program, "NASA Fundamental Aeronautics Program," presentation to the committee, November 19, 2009, pp. 19 and 30.

SAFETY-RELATED RESEARCH IN THE AIRSPACE SYSTEMS PROGRAM

The committee also learned that some aviation safety-related research is conducted within the Airspace Systems Program, in addition to the Aviation Safety Program and the Fundamental Aeronautics Program. Unfortunately, as with the research in the Fundamental Aeronautics Program, the committee could not delve into the research as thoroughly as it was able to examine the research in the Aviation Safety Program; thus its assessment is at a somewhat higher level than the six research concerns described earlier in the chapter.

Defining and Prioritizing Research Objectives

The goal of the Airspace Systems Program is to "develop and demonstrate future concepts, capabilities, and technologies that will enable major increases in the air traffic management capacity, flexibility, and efficiency, while maintaining safety, to meet NextGen requirements."[29] As with research in the Fundamental Aeronautics Program, improving aviation safety is not the primary concern for the Airspace Systems Program. Research within the program is divided into two primary project areas:

- The NextGen-Airspace Project, which encompasses a wide variety of topics broadly dealing with en route, transition, and terminal area airspace, and
- The NextGen Airportal Project, which covers low altitude terminal area airspace and the airport environment.

NASA briefers identified research within the Airspace Systems Program that has a strong capacity and safety focus:

- Separation assurance for en route and transition airspace and airport surface,
- Terminal area operations, and
- Wake vortex.

NASA briefers pointed to several of the mobility goals from the National Plan as the rationale behind these research areas, particularly mobility goal 1, "Develop reduced aircraft separation in trajectory- and performance-based operations." NASA briefers did not cite relevant challenges from the NRC *Decadal Survey,* but did point to NASA's 2006 strategic plan as additional reasoning behind the safety research, particularly outcome 3E.2:

> By 2016, develop and demonstrate future concepts, capabilities, and technologies that will enable major increases in air traffic management effectiveness, flexibility, and efficiency, while maintaining safety, to meet capacity and mobility requirements of the Next Generation Air Transportation System.[30]

Additionally, briefers pointed to two JPDO high-priority focus areas that the Airspace Systems Program is involved in: supporting the Aviation Safety Program's development of V&V and leading research in increasing the "clarity of air/ground functional allocation"[31] (a JPDO initiative to develop a decision roadmap for the evolution of roles for the flight deck, air traffic controller, and automation, including associated operations changes).

Overall, the committee observed that the safety research being conducted by the Airspace Systems Program appears fairly well defined and appropriate. Whether these research activities are correctly prioritized, however, was not a question the committee could answer. The committee's concerns regarding V&V, as discussed earlier in the chapter, still apply. There appears to be active coordination between this program's researchers and those in the Aviation Safety Program in several areas, although the efforts appear to be more focused on maintaining

[29] John Cavolowsky, Director, Airspace Systems Program, "Airspace Systems Program," presentation to the committee, November 19, 2009, p. 3.

[30] John Cavolowsky, Director, Airspace Systems Program, "Airspace Systems Program," presentation to the committee, November 19, 2009, p. 12.

[31] John Cavolowsky, Director, Airspace Systems Program, "Airspace Systems Program," presentation to the committee, November 19, 2009, p. 14.

the level of safety within the system as opposed to improving safety. Given the similarity in general objectives, the committee expected to find a large amount of coordination between the Airspace Systems Program and the Aviation Safety Program in New Operations research. However, the committee observed a relatively low amount of coordination.

Resource Allocation

Unlike other briefings to the committee, presentations on the Airspace Systems Program did not specifically indicate the amount of resources being allocated to its safety-related research. This may be because the safety research is interspersed among many different subprojects. Figures 3.10 and 3.11 show the resource allocation information shared by NASA briefers with the committee.

Even if the briefers had separated the data to show only the safety-related research, it likely would have been difficult or impossible for the committee to assess the appropriateness of the allocation, for the same reasons as described earlier in the chapter for other research concerns.

Coordination with the FAA and Others

NASA briefers indicated that the safety research in the Airspace Systems Program is coordinated with a variety of government entities through interagency agreements, including the FAA, NOAA, the Department of Transportation, and the USAF. Briefers reported a high degree of alignment between Airspace Systems Program milestones and the JPDO's research and development needs. The program has several Space Act agreements in place with private companies, such as Lockheed Martin and Boeing, as well. Overall, the committee observed a high degree of research coordination with non-NASA entities. The committee did not have the chance to delve into the specifics of these efforts, however, so the committee was unable to assess the actual depth of the program's coordination.

Research Transition

The committee was very impressed with the research transition mechanisms utilized by the Airspace Systems Program. Briefers gave in-depth descriptions of several Research Transition Teams (RTTs), groups that were developed to "ensure that R&D needed for NextGen implementation is identified, conducted, and effectively transitioned to the implementing agency."[32] These and other transition efforts span numerous technologies being developed within the program applicable to a variety of areas, including:

- Dynamic Airspace Configuration,
- Efficient Flow into Congested Airspace,
- Integrated Arrival/Departure/Surface, and
- Flow-Based Trajectory Management.

Overall, the committee observed that these activities appear to be suitable for timely transition of research results.

SUMMARY ASSESSMENT

The committee was charged with examining NASA's aviation safety-related research. The bulk of this research is conducted in ARMD's Aviation Safety Program, and managers from this program split the research into six primary concerns. Additional relevant research is conducted in ARMD's Fundamental Aeronautics Program and Airspace Systems Program. The committee examined all of this research and strove to answer four key questions:

[32] John Cavolowsky, Director, Airspace Systems Program, "Airspace Systems Program," presentation to the committee, November 19, 2009, p. 27.

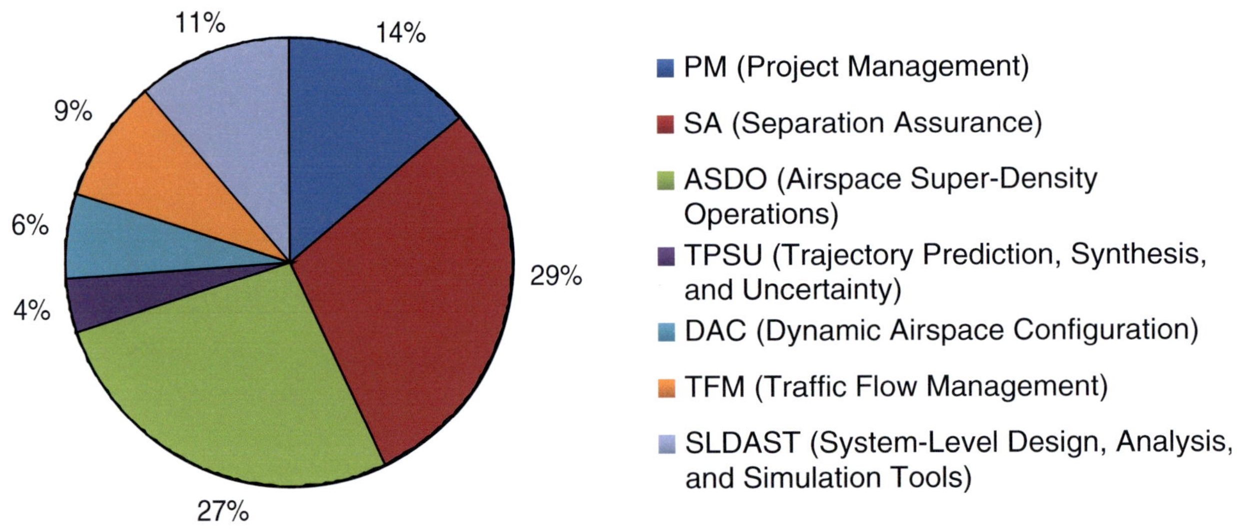

FIGURE 3.10 Airspace Systems Program's FY 2009 budget. SOURCE: John Cavolowsky, Director, Airspace Systems Program, "Airspace Systems Program," presentation to the committee, November 19, 2009, p. 30.

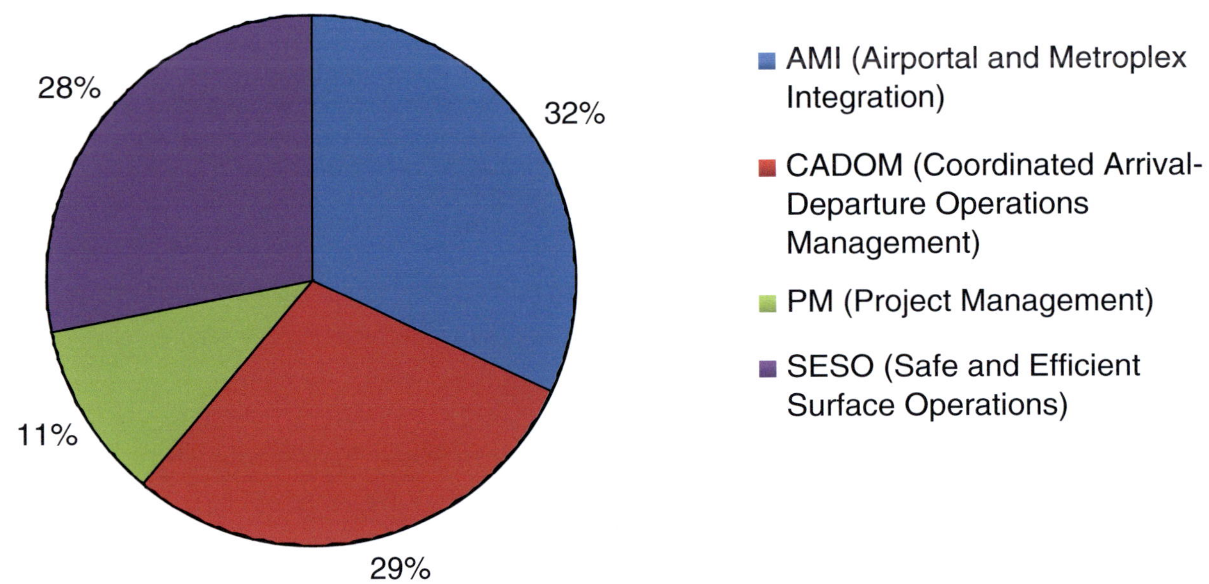

FIGURE 3.11 FY 2009 Budget of the Airportal Project. SOURCE: Airspace Systems Program's FY 2009 budget. SOURCE: John Cavolowsky, Director, Airspace Systems Program, "Airspace Systems Program," presentation to the committee, November 19, 2009, p. 30.

1. Are the research objectives well defined, prioritized, and appropriate?
2. Is the research properly coordinated with the research programs of the FAA and other relevant federal agencies?
3. Are appropriate resources allocated for the research?
4. Do suitable mechanisms exist for transitioning the research results into operational technologies and procedures in a timely manner?

Defining and Prioritizing the Research Objectives

The committee observed deficiencies with regard to how NASA defines and prioritizes specific research activities across all six research concerns. It was difficult for the committee to understand how NASA prioritized objectives and activities within each research concern, due to inadequate discussion of the matter in NASA's technical plans and briefings to the committee. In some cases, the committee was not presented with a strong rationale for NASA's chosen priorities, such as the emphasis on commercial air transport in New Operations. The committee also observed a tendency for prioritization based on existing resources and personnel; this was particularly the case in Aircraft Aging and Durability and was evidenced elsewhere, such as in the coordination of areas of historical expertise in On-Board System Failures and Faults. In order to advance the state of aviation safety, one of ARMD's core missions, it is incumbent upon ARMD to expand its capabilities beyond its traditional areas of expertise to meet the requirements of emerging and future national safety priorities.

In addition, the committee questions some of NASA's chosen priorities, which result in, for example, insufficient attention being given to human factors and automation in Loss of Control research. ARMD has chosen to focus on longer-term, fundamental research, with which the committee agrees. However, research in aviation safety must have a path to implementation. The committee could not always see this path for some research projects, and the individual NASA researchers could not provide it.

The committee also observed less coordination within NASA for the research concerns than it expected to find, particularly between Loss of Control and New Operations. The committee also expected, but did not observe, additional coordination between researchers in New Operations and the Airspace Systems Program.

Resource Allocation

As with prioritization of research activities, resource allocation was generally difficult to assess for the committee. The committee observed several instances of questionable allocation, such as the emphasis on adaptive control in Loss of Control research or the relatively low amount of research being conducted on information management and processing compared to the amount of funds being directed into it. However, these were generally exceptions rather than the rule, as much of the resource allocation in the other research concerns appeared appropriate under the committee's high-level review. Briefings to the committee by NASA, and ARMD's various technical plans, lacked the necessary details that would enable more specific findings from the committee. The committee has some concerns regarding NASA's commitment to developing a core competency in V&V based on the variation in resources over the years and the large amount of research being conducted through NRAs. For V&V to become another node of high-quality research for NASA, the research must be conducted primarily in-house and have stable funding; this may be the case if the president's FY 2011 budget passes.

Coordination with the FAA and Others

The committee notes that NASA has done a good job, and even excelled in some instances, in coordinating its research with the FAA and other relevant federal agencies. The level of collaboration observed in HIWC activities, part of the Hazardous Conditions research concern, could be viewed as a prime model of how NASA can keep its research relevant to the needs of others within the National Airspace System. The depth and intensity of coordination in some of the other research concerns were difficult for the committee to judge, but in general it appears that coordination of aviation safety-related research is occurring and is being actively pursued by NASA. Within several of the research concerns, particularly in Analyzing Complex Systems for Safety and in On-Board System Failures and Faults, NASA has distinct opportunities to enhance its coordination efforts.

Research Transition

As with its coordination activities, NASA generally appears to have strong mechanisms in place to transition its research results. However, the committee notes that it did not observe any specific mechanisms or examples of technology transition in the On-Board System Failures and Faults research concern, and the examples of transitioning provided by NASA in the Analyzing Complex Systems for Safety research area were questionable at best. However, improvements in this latter area are likely in the future as NASA's V&V research starts in earnest. In reviewing NASA's transition mechanisms, the committee identified an additional area of possible research for NASA—certification of non-deterministic software.

The committee also observes that ARMD has chosen to focus on longer-term, fundamental research, with which the committee agrees. However, research in aviation safety must have a path to implementation. The committee could not always see this path for some research projects, and the individual NASA researchers could not provide it. Additional discussion of NASA's possible role in overcoming other certification and implementation barriers is discussed in the next chapter.

4

Key Findings and Recommendations

Advancing the state of aviation safety is central to the mission of the National Aeronautics and Space Administration (NASA) and its Aeronautics Research Mission Directorate (ARMD). In keeping with direction from the *National Aeronautics Research and Development Policy*[1] (referred to as the "National Policy") and the *National Plan for Aeronautics Research and Development and Related Infrastructure*)[2] (referred to as the "National Plan"), ARMD's emphasis is on pursuing what it describes as "long-term, cutting-edge research that expands the boundaries of aeronautical knowledge for the benefit of the broad aeronautics community." In particular, ARMD is charged with undertaking "foundational" research that consists of basic research and developing a strong aeronautics knowledge and technology base to overcome barriers to technological progress in aviation.

Much of ARMD's safety-related research is aimed at predicting and preventing safety hazards in an increasingly diverse and complex Next Generation Air Transportation System (NextGen). Underlying this interest is concern that as NextGen capabilities are introduced and the national airspace becomes more heavily used by a diversity of aircraft, new safety hazards will arise that require innovative solutions through research. Yet, not only must these future safety hazards be predicted and prevented with NASA's help, but the agency also will continue to be called on to find solutions to safety problems arising in the current system. While this system performs at very high levels of safety, there is an expectation that safety advancements will continue, both in the near and longer terms.

Congress requested this review of NASA's aviation safety-related research programs. This chapter presents the main findings of the review as they relate to the four questions in the congressional request. Some of the findings go beyond these specific queries. For instance, in assessing NASA's safety research coordination and transitioning activities, consideration is given to the connections established with industry, academia, and the aviation community generally as well as those forged with the Federal Aviation Administration (FAA) and other federal agencies. Because NASA's aeronautics research constitutes a key component of the nation's aviation research enterprise, its safety-related activities warrant consideration within this larger context.

---

[1] National Science and Technology Council, *National Aeronautics Research and Development Policy,* Office of Science and Technology Policy, Executive Office of the President, Washington, D.C., December 2006, available at http://www.aeronautics.nasa.gov/releases/national_aeronautics_rd_policy_dec_2006.pdf.

[2] National Science and Technology Council, *National Plan for Aeronautics Research and Development and Related Infrastructure,* Office of Science and Technology Policy, Executive Office of the President, Washington, D.C., December 2007, available at http://www.aeronautics.nasa.gov/releases/aero_rd_plan_final_21_dec_2007.pdf.

The findings indicate that NASA's aeronautics research enterprise has made, and continues to make, valuable contributions to aviation system safety, but it is falling short in some key respects. To address these shortcomings, the committee offers several recommendations at the conclusion of this chapter.

KEY FINDINGS

Do NASA's Safety-Related Research Programs Have Well-Defined, Prioritized, and Appropriate Research Objectives?

The National Plan identifies a series of fundamental research challenges that are top national priorities and are intended to provide strategic guidance for all federal civil aeronautics research and development (R&D). In accordance with the plan's challenges for safety research, ARMD's Aviation Safety Program has established research projects to address the following concerns:

- New Operations,
- Flight In or Around Hazardous Conditions,
- Loss of Control,
- Durable Aircraft Structures and Systems,
- On-Board System Failures and Faults, and
- Analyzing Complex Systems for Safety.

As guideposts for programming safety research, these appear to be worthwhile objectives, consistent with the challenges identified in the National Plan and reflective of current and emerging safety issues. Whether the research being undertaken by ARMD to further these six objectives is sufficiently defined, prioritized, and appropriate, however, is a separate matter.

Because there are many aspects to each of the six research objectives, ARMD could choose to pursue any number of potential research topics commensurate with each. To make the best use of its resources requires that ARMD carefully examine the research needs associated with each safety objective and determine where it can contribute the most to meeting these needs in light of its own research competencies and work being undertaken elsewhere. The committee, therefore, expected to find a research prioritization process that is deliberate and well informed—supported by empirical analyses and advice from experts—and accompanied by a well-documented rationale for the program content.

In not finding a prioritization process resembling this ideal, the committee struggled in understanding why ARMD safety research focuses on certain issues and topics while paying little, if any, attention to others. The committee observed, for instance, that much of the research addressing "loss of control" centers on automation by seeking to advance the state of the art of adaptive control systems for aircraft. The basis for this research interest—as opposed to research on other topics relevant to loss of control, such as human interactions with automation—was neither obvious nor well explained. More generally, the committee observed an emphasis throughout the research programs on safety issues that apply to commercial air transport but relatively little work having direct pertinence to unmanned aircraft or general aviation, even though the former represents a potential vehicle configuration in NextGen and the latter has long accounted for the majority of fatal aviation crashes.

By not having a defensible, analytically based process for prioritizing its safety research, ARMD could not justify, in a convincing manner, much of the content of its research programs. Thus, in not having access to such an independent assessment of safety research needs, the committee could not determine whether ARMD's safety research programs are prioritized to make the best use of available resources and neither can ARMD.

Finding 1: *NASA needs a more objective process for prioritizing safety research.* While the objectives of ARMD's Aviation Safety Program are worthy guideposts for safety research, ARMD lacks a well-founded process for prioritizing the research needs associated with each objective, and thus for ensuring that its research is well aligned with meeting critical national aviation safety needs.

Although not possessing the qualities defined above, ARMD does employ a process for prioritizing and programming its safety research in accordance with its safety objectives. Based on the committee's review of the research content and ARMD's own description of its programming methods, this process appears to be driven largely by ARMD's interest in employing existing personnel and assets at the NASA research centers. Those safety objectives that map well with ongoing research activities and with these internal interests are generally given priority in the programming of research and allocation of resources. One example (discussed in Chapter 3) that illustrates this observation is the selection of research challenges in the Aircraft Aging and Durability Project that align with "NASA's core capabilities."[3]

This internally driven process is undoubtedly a manifestation of a number of constraints and demands on ARMD, which the committee could not fully explore. NASA's aeronautics research enterprise has experienced budgetary retraction for more than a decade, during which time it has undergone multiple reorganizations and changes in leadership, mission, and goals. These factors, among others, may have contributed to a climate that emphasizes the protection and preservation of existing research activities, personnel, and assets. Whether this method of research programming leads to a marginally or fundamentally different safety research portfolio than would have emerged using a more objective process cannot be readily determined. It is self-evident, however, that any prioritization process driven by an organization's existing interests will risk neglecting new and emerging safety needs that are not well aligned with the status quo.

Finding 2: *Internal interests are overemphasized in the programming of safety research.* ARMD gives undue weight to research that aligns well with its existing activities, personnel, and assets rather than the results of critical evaluations of current and emerging aviation safety needs.

Have Resources Been Allocated Appropriately to Research Objectives?

ARMD currently has research competencies that are recognized nationally and internationally in critical safety areas, such as icing research. While these existing nodes of expertise need to be recognized and their critical mass sustained, they risk being neglected as research funding is spread widely to preserve all of ARMD's research capabilities, including those that are no longer unique to NASA or of high safety relevance. Indeed, the committee suspects ARMD has already lost this critical mass in the important safety area of human factors.

Yet, even as ARMD seeks to retain and strengthen its core safety research competencies, it must give sufficient attention to investing in the research expertise that will be needed to address new and emerging safety issues. ARMD recognizes the importance of such forward-looking investments, as evidenced by its efforts to expand and strengthen its expertise in software verification and validation (V&V). In relying too much on program budgetary increases to fund such investments, however, ARMD may be seriously hampered in its ability to make such critical investments in its workforce and facilities. To acquire the needed expertise going forward, ARMD will almost certainly need to make many difficult decisions about the allocation of resources among its existing facilities and program areas, requiring that some capabilities be eliminated and others substantially scaled back. Not having in place an objective and well-informed means of assessing safety needs and priorities makes building the case for such resource investments and realignments even more difficult.

Finding 3: *Too few resources are devoted to sustaining and acquiring critical safety research capabilities.* Continued emphasis on preserving existing research expertise and assets risks degradation of ARMD's core safety research strengths and the prolonged neglect of competencies required to address new and emerging safety issues.

In light of ARMD's emphasis on advanced and long-range research, the committee is surprised to find few programmatic means and resources set aside for fostering exploratory and innovative thinking on ways to solve

[3] NASA, *Aviation Safety Program Aircraft Aging and Durability Project Technical Plan Summary,* Washington, D.C., available at http://www.aeronautics.nasa.gov/nra_pdf/aad_technical_plan_c1.pdf, p. 2.

safety problems. Having mechanisms to invite such innovation would seem to be an important aspect of a research enterprise intent on engaging in "cutting-edge research that expands the boundaries of aeronautical knowledge." Even though funding such research may involve greater uncertainty about expected payoffs, taking such calculated risks may be warranted in cases where research investments to solve safety problems have reached the point of diminishing returns and for which breakthrough insights and technologies are needed.

Most of ARMD's aviation safety work is performed by agency personnel and contractors at NASA research centers. Less than 15 percent of the funding for the Aviation Safety Program goes to outside researchers through the NASA Research Announcements (NRA). For the most part, ARMD uses the NRA program to sponsor external research that is closely aligned with its existing internal projects, and thus to make incremental contributions to them. Used in this way, NRA's provide little incentive or latitude for researchers from academia, industry, and independent research institutes—as well as from within NASA—to propose and explore new and innovative ideas to enhance safety. Moreover, ARMD does not have any other formal mechanisms in place to support such exploratory research and innovation.

Finding 4: *Too few resources and programs are devoted to stimulating innovation.* ARMD lacks the structure to elicit, explore, and develop innovative ideas to advance aviation safety.

Are the Programs Properly Coordinated with the Safety Research Programs of FAA and Other Relevant Federal Agencies?

At the agency level, NASA, FAA, and other federal agencies collaborate in developing the National Plan, which lays out the fundamental safety challenges that are a priority for the entire federal civil aviation research enterprise. At the program level, ARMD and its Aviation Safety Program also have strong connections to the Joint Planning and Development Office (JPDO), enabling a stronger understanding of research needs associated with NextGen and NASA's role in meeting them. At the project level, NASA and FAA coordinate during the conduct of safety research, often quite effectively and with many positive interactions as evidenced by several of the projects reviewed by the committee. Indeed, an outstanding example of this collaboration is the work addressing the issue of high ice water content (HIWC) hazards.

Such varied means of coordination across federal agencies are commendable and critical.

Yet, while much of NASA's safety-related research is relevant to the needs of the FAA, much of it also pertains to the aviation community more generally. In this regard, NASA's connections must extend beyond the FAA to include industry and academia. The interconnectedness and complexity of the aviation system demands that NASA seek the broadest possible cooperation and coordination of all parties involved. Indeed, the committee finds that ARMD has continuing involvement and interactions with a broad array of safety-related industry-led groups, such as the Commercial Aviation Safety Team (CAST), the Aviation Safety Information Analysis and Sharing system (ASIAS), and many discipline- and problem-oriented groups, such as the Joint Aircraft Survivability Program and Joint Council on Aging Aircraft. While it is difficult to know the intensity of these interactions based on NASA briefings and committee reviews of planning documents, there seem to be many avenues for NASA interaction and coordination across the aviation community.

One area where greater inclusion is desirable is in seeking external reviews of ARMD safety research activities. Aviation Safety Program activities are currently reviewed on a periodic basis by experts from the FAA and other federal agencies. Such agency reviews and consultations are vital to ensuring the research is compatible with the work going on elsewhere in the federal government and is relevant to the operational and regulatory needs of federal agencies. Limiting these external reviews to experts from federal agencies, however, may cause ARMD to miss out on opportunities to gain additional perspective from industry and academia and to inform outside perceptions of the program's capabilities and priorities.

There is a general recognition within the research community that external reviews foster higher-quality research. Yet, it is important that the reviews be conducted not only at the end stages of the research but on a continuing basis, from the beginning of the work to its completion and transitioning. Seeking such advice early in the research process is important because this is when research managers have the greatest latitude to reshape the program. Interim reviews as the work progresses are helpful for ensuring that the work remains of high quality and retains safety relevance. If the research holds sufficient promise, some of the reviewers themselves may wish to partner in the work; after the work is completed, the reviewers may become early users and champions of the results.

There are a variety of mechanisms for obtaining such continuous input, such as special project review committees, advisory committees, ad hoc review teams, and individual reviews by recognized experts. Because a good portion of ARMD's research is not intended to be near term and highly applied in nature, it would be important to ensure that external advisors recognize this and consist of some individuals with an understanding of broader safety-related trends and technology development processes.

ARMD sponsors many activities aimed at information dissemination and exchange. The Aviation Safety Program convenes an annual Safety Technical Conference and hosts other outreach activities, such as the Airspace Systems Program's Technical Exchange Meeting and NextGen Workshops. These forums provide valuable opportunities for researchers to present results and progress and for the exchange of research ideas and technical information. While the kind of external reviews described above would have a different purpose than these outreach activities, they would be highly complementary.

Finding 5: *Connections with the FAA, other federal agencies, and the aviation community are varied but not deep.* NASA and the FAA coordinate in the planning and conduct of safety research, and many mechanisms exist for interacting and exchanging information with other federal agencies, academia, and industry. These connections could be deepened through more inclusive and sustained reviews of NASA safety research by such outside experts.

Safety research is the main mission of ARMD's Aviation Safety Program. While having a program that is dedicated to aviation safety is important, a single program cannot be expected to cover the full gamut of aviation safety issues. In the large, evolving, and complex aviation system, safety concerns can arise from interactions among system elements that may be considered safe on their own but not in combination, such as interactions among aircraft and airspace technologies and procedures. Under these circumstances, one would expect to find a significant amount of safety-related research being planned and undertaken in the other ARMD programs, often in collaboration with the Aviation Safety Program. With a few exceptions, however, ARMD's safety research programs were presented to the committee as discrete activities of the Aviation Safety Program, the Fundamental Aeronautics Program, and the Airspace Systems Program.

Yet in reviewing each of these research programs, it proved difficult to identify the full array of safety-related research in the Fundamental Aeronautics Program and the Airspace Systems Program. One reason for this difficulty is that the research in these programs is often characterized as focusing on other objectives, such as increasing operational efficiency and developing innovative aircraft designs, structures, and materials. In such cases, it appears as though safety is treated more like a constraint than an opportunity—that is, safety performance is viewed as something the research cannot diminish, rather than something that it can improve.

Given the complexity and breadth of the aviation safety challenge, it is troubling to find such "stove-piping," or compartmentalization, within ARMD. Even within the Aviation Safety Program itself, however, the committee observes this phenomenon, such as the relatively few collaborative activities occurring across the related objective areas of New Operations and Loss of Control and across the Aircraft Aging and Durability and Integrated Vehicle Health Management projects.

Finding 6: *Internal coordination of and collaboration on safety research need improvement.* Within ARMD, there is stove-piping of research that risks system-level safety solutions not being explored and safety hazards not being addressed that arise from interactions among aviation system elements.

Do Suitable Mechanisms Exist for Transitioning the Research Results from the Programs into Operational Technologies and Procedures and Certification Activities in a Timely Manner?

Exploiting the opportunities described above for coordination and collaboration with the FAA and industry can be critical to the successful transitioning of research results to follow-on stages of development or application. The aforementioned HIWC activity exemplifies how such coordination and collaboration can foster the transfer of research results to the field. In reviewing the Aviation Safety Program's research portfolios, the committee observed many instances of researchers partnering with industry through use of Space Act agreements and other mechanisms designed to build such connections with end-users and ensure the research is relevant and remains on a productive course. At the same time, however, there are many practical challenges to building and maintaining such connections in light of ARMD's emphasis on longer-range research and the demands of the FAA and industry for relatively mature and well-tested products and procedures.

Because much of ARMD's safety work consists of fundamental and long-term research, a high degree of operational applicability cannot be a metric for judging whether the research should be undertaken, nor can there be a strong emphasis on definitive end products and transition planning. Yet, for some of the safety-relevant research being pursued by ARMD, the resulting technologies and process are intended to support FAA operational needs and have application, in which case transition planning is important. A prominent example is ARMD's work on air traffic management (ATM) technologies and procedures, such as trajectory-based operations and weather avoidance. The utility of NASA's work in this area has long been questioned because of the high levels of safety assurance demanded by the FAA when making changes to ATM technologies and procedures. The committee observes, however, that NASA and the FAA recognize these challenges and are actively seeking to overcome them, by for instance, coordinating special research transition teams intent on identifying and addressing safety implications much earlier in the research process.

Even for research that is longer-term in nature, NASA must have some understanding of eventual implementation challenges in order to judge the merits of the work. The Aviation Safety Program's efforts on advancing the state of the art of adaptive systems for loss of control is a clear example of this need. This is because achieving an accepted level of safety assurance required for certification of the non-deterministic software used in these systems promises to be a difficult and prolonged undertaking. In this case, the implementation challenge itself may be more complicated and daunting than the development of the control technologies themselves.

The committee believes the kind of external reviews espoused above are likely to bring such implementation challenges to the forefront, compelling early attention by those programming and engaging in the research. By applying foresight in mapping out the implementation challenges, researchers can better assess the long-term practicality of the research in relation to other research options addressing the same safety objective. Doing so will not only yield insight into these implementation challenges, but may suggest areas where ARMD research can be helpful in overcoming them, such as in developing improved safety analysis and approval methods.

> **Finding 7:** *Demands for safety-assured technologies and procedures can conflict with NASA's emphasis on long-range, foundational research.* ARMD exploits many mechanisms to assist in furthering the technologies and procedures developed through its research; however, safety assurance and approval requirements can present vexing implementation challenges. In light of these challenges, some of ARMD's safety-related research would appear to have very limited prospects for eventual implementation—a risk that deserves more explicit consideration when ARMD programs its research.

RECOMMENDATIONS

Recommendation 1: The Aeronautics Research Mission Directorate (ARMD) should adopt a more fully informed, empirical, and documented process for identifying and prioritizing safety research needs for use in guiding its aeronautics research and development programming and investments in research expertise and capacity. A central element of this process should be the development of comprehensive aviation safety needs assessments. These assessments should be:

- Made objectively, independent of ARMD's existing research expertise, assets, and resource requirements and constraints.
- Undertaken in close coordination with the Federal Aviation Administration and other relevant federal agencies and in consultation with industry, academia, other safety-related organizations, and considering the relevant aviation safety data and literature.
- Cognizant of the safety needs being researched elsewhere in government, industry, and academia to know where critical gaps in research coverage may exist.

Recommendation 2: The Aeronautics Research Mission Directorate should establish programmatic means for encouraging more exploratory research on innovative ideas to improve aviation safety. The program should elicit and develop the promising ideas of researchers from industry, academia, other government agencies, and NASA.

Recommendation 3: The Aeronautics Research Mission Directorate's safety-related research activities should be subject to regular reviews by outside experts from the Federal Aviation Administration and other government agencies, industry, independent research institutes, and universities. These reviews, which will help in ensuring continued safety relevance, quality, implementation challenges, and successful transitioning, should be undertaken during the formative stages of the research, interim phases, and as the work is being completed. The reviews should have a prominent role in informing research programming decisions.

Recommendation 4: The Aeronautics Research Mission Directorate should develop and implement processes that will lead to more coordination and collaboration in the planning and conduct of safety research both within the Aviation Safety Program and across all its aeronautics research programs.

Appendixes

A

Statement of Task

The National Research Council's Aeronautics and Space Engineering Board, in conjunction with the Transportation Research Board, will establish an ad hoc study committee to conduct an independent review of NASA's aviation safety-related research programs. The review shall assess whether:

(1) The programs have well-defined, prioritized, and appropriate research objectives;
(2) The programs are properly coordinated with the safety research programs of the Federal Aviation Administration and other relevant federal agencies;
(3) The programs have allocated appropriate resources to each of the research objectives; and
(4) Suitable mechanisms exist for transitioning the research results from the programs into operational technologies and procedures and certification activities in a timely manner.

B

Committee Biographical Information

H. NORMAN ABRAMSON, *Chair,* is executive vice president (emeritus) of Southwest Research Institute. He is internationally known in the field of theoretical and applied mechanics. His specific area of expertise is in the dynamics of contained liquids in astronautical, nuclear, and marine systems. He began his career as an associate professor of aeronautical engineering at Texas A&M University and has served as vice president and governor of the American Society of Mechanical Engineers (ASME) and director of the American Institute of Aeronautics and Astronautics (AIAA). He is an AIAA fellow and fellow and honorary member of the ASME. He received his Ph.D. in engineering mechanics from the University of Texas at Austin. As a member of the National Academy of Engineering, he served on its council from 1984 to 1990. He has also chaired and served on many other NAE and NRC committees, including chair of the Committee on R&D Strategies to Improve Surface Transportation Security, the Transportation Research Board's (TRB's) Research and Technology Coordinating Committee, and TRB's Federal Transportation R&D Strategic Planning Process. He was as a member of the U.S. Air Force Scientific Advisory Board from 1986 to 1990. From 2000 to 2002, he chaired the Committee for a Study of Public Sector Requirements for a Small Aircraft Transportation System, which was sponsored by NASA. He was vice chair of the Committee for a Review of the National Transportation Science and Technology Strategy. He recently served on the Committee on National Institute of Aerospace Proposal Reviews and the Committee on the Role of Naval Forces in the Global War on Terror. He is currently a member of the Oversight Committee for the Strategic Highway Research Program 2.

ROBERT T. FRANCIS, *Vice Chair,* is senior policy advisor for Zuckert, Scoutt, and Rasenberger, LLP. He is the past vice chair of the National Transportation Safety Board (NTSB). After joining the NTSB, he was the senior official at a number of transportation accident investigations, including the explosion and crash of TWA Flight 800 off Long Island, New York; the crash of ValuJet Flight 592 in the Florida Everglades; and a Learjet 35 accident in Mina, South Dakota. Mr. Francis also has chaired a number of NTSB public hearings, including the hearing on Part 145 aviation maintenance practices and oversight, the hearing on Korean Air Flight 801 that crashed in Guam, and the hearing on passive grade crossing safety in the United States. Prior to his appointment to NTSB, he served as senior representative for the Federal Aviation Administration (FAA) in western Europe and North Africa. Representing the FAA administrator, he worked extensively on aviation safety and security issues with U.S. and foreign air carriers, transportation governmental authorities, aircraft manufacturers, and airports. He is a member of the Executive Committee of the Board of Governors of the Flight Safety Foundation and is actively involved as a member of the foundation's ICARUS Committee, a group of worldwide aviation experts who gather informally to share ideas on reducing human error in the cockpit. He has specialized in international aviation safety issues and has spoken extensively on this subject. Mr. Francis is a recipient of an Aviation Week and Space Technology 1996 Laurels Award and was recognized by both the U.S. Navy and the U.S. Coast Guard for meritorious service in the TWA Flight 800 investigation. He received his A.B. from Williams College.

ELLA M. ATKINS is associate professor of aeronautical engineering, University of Michigan. Her research focuses on the integration of strategic and tactical planning and optimization algorithms to enable robust, autonomous aircraft and spacecraft flight in the presence of system failures and environmental uncertainties. Before joining the faculty of the University of Michigan, she was assistant professor of aerospace engineering at the University of Maryland. She is author of more than 60 archival and conference publications and serves as an associate editor for the AIAA *Journal of Aerospace Computing, Information, and Communication*. She is a technical program chair for the AIAA Infotech@Aerospace conference and has served on several review boards and panels. She is chair-elect of the AIAA Intelligent Systems Technical Committee, an associate fellow of AIAA, a member of the Association for the Advancement of Artificial Intelligence and the Institute of Electrical and Electronics Engineers (IEEE), and a private pilot. She earned her B.S., M.S, and Ph.D. in aeronautics and astronautics from the Massachusetts Institute of Technology (MIT) and an M.S. and Ph.D. in computer science and engineering from the University of Michigan. She has served on the NRC's Panel E: Intelligent and Autonomous Systems, Operations and Decision Making, Human Integrated Systems, Networking, and Communications.

DEBORAH A. BOEHM-DAVIS is university professor and chair of the psychology department at George Mason University. She worked on applied cognitive research at General Electric, NASA Ames Research Center, and Bell Laboratories prior to joining George Mason University in 1984. She is also the recipient of a Medical Devices Fellowship Program award that allowed her to serve as a senior policy advisor for human factors at the Food and Drug Administration's Center for Devices and Radiological Health. Her research interests include the analysis of pilot procedures and practices for automated flight decks. She has served as president of Division 21 (Applied Experimental and Engineering Psychology) of the American Psychological Association (APA) and as president and secretary-treasurer of the Human Factors and Ergonomics Society (HFES). She is an associate editor for *Human Factors* and the *International Journal of Human-Computer Studies,* and she serves on the editorial board of *Theoretical Issues in Ergonomic Sciences*. In 2003, she received the Franklin V. Taylor Award from Division 21 of the American Psychological Association. She is a fellow of the APA, the HFES, and the International Ergonomics Association. She holds an A.B. in psychology from Rutgers the State University (Douglass College) and an M.A. and Ph.D. in cognitive psychology from the University of California, Berkeley. Dr. Boehm-Davis is currently a member of the NRC Committee on Human-Systems Integration.

JAMES BURIN is director of technical programs at the Flight Safety Foundation (FSF), an independent, nonprofit organization engaged in research, education, and advocacy to improve aviation safety. As director, Mr. Burin organizes and oversees safety committees and manages safety-related conferences and research. Prior to joining the FSF, he was the director of the School of Aviation Safety, Monterey, California. He has 40 years of aviation experience and 32 years of experience in the field of aviation safety. He is a retired Navy captain, having commanded an attack squadron and carrier air wing during his 30-year military career. His work on aviation safety includes controlled flight into terrain, human factors, safety program organization, accident investigation, operations, education, and organizational and leadership influences on safety. He earned a B.A. from Dartmouth College and an M.S. in systems analysis from the Naval Postgraduate School.

COLIN G. DRURY is distinguished professor emeritus and chair of industrial engineering of the State of University of New York at Buffalo, concentrating on the application of human factors techniques to manufacturing and maintenance processes. He was manager of ergonomics at Pilkington Glass. He has published extensively on topics in industrial process control, quality control, aviation maintenance, and safety and was the North American editor of *Applied Ergonomics*. From 1988 to 1993, he was the founding executive director of the Center for Industrial

Effectiveness. He is a fellow of the Institute of Industrial Engineers, the Ergonomics Society, the Human Factors Ergonomics Society (HFES), and the International Ergonomics Society. Dr. Drury received the Bartlett Medal of the Ergonomics Society and the Fitts Award of the Human Factors Ergonomics Society. He was also awarded the HFES's Lauer Safety Award, the FAA's Excellence in Aviation Research Award, and the American Association of Engineering Society's Kenneth A. Roe Award. He received his B.Sc. in physics from the University of Sheffield and his Ph.D. in engineering production from the University of Birmingham, United Kingdom. He has served on a number of NRC committees including the Committee on Assessment of Security Technologies for Transportation and the Committee to Evaluate the Safety and Environmental Metrics for Potential Application at Chemical Agent Disposal Facilities.

R. JOHN HANSMAN, JR. is a professor and director of the MIT International Center for Air Transportation. In addition to teaching, Dr. Hansman conducts research in several areas related to air transportation, flight vehicle operations, and safety. His current research activities focus on information technology applied to air transportation systems, air traffic control, integrated human-automation systems, advanced vehicles, and advanced cockpit information systems. He is also an internationally recognized expert in aviation meteorological hazards, such as icing and wind shear. He received his Ph.D. in physics, meteorology, aeronautics, and astronautics from MIT. Dr. Hansman served on the NRC Committee to Identify Potential Breakthrough Technologies and Assess Long-Term R&D Goals in Aeronautics and Space Transportation Technology and the Committee on the Effects of Aircraft-Pilot Coupling on Flight Safety.

PIERRE T. KABAMBA is a professor of aerospace engineering in the Department of Aerospace Engineering at the University of Michigan. His research interests are in the area of linear and nonlinear dynamic systems, robust control, guidance and navigation, and intelligent control. His recent research activities have aimed at the development of quasi-linear control theory and the design, scheduling, and operation of multi-spacecraft interferometric imaging systems to be used to obtain images of exo-solar planets. Moreover, he has done work in the analysis and optimization of random search algorithms. He is also doing work in simultaneous path planning and communication scheduling for Unmanned Aerial Vehicles (UAVs) under the constraint of radar avoidance. He is author or co-author of more than 170 publications in refereed journals and refereed conferences. He is a fellow of the IEEE. He received his Ph.D. in mechanical engineering from Columbia University.

JAMES R. KRODEL is a fellow of the Control Systems Verification and Validation Group at Pratt and Whitney Jet Engines. Mr. Krodel has more than 30 years of experience in the aerospace software domain. He has held several technical and managerial positions in software development of embedded systems, including software for the full authority digital electronic engine control for the Pratt and Whitney PW4084 jet engine propulsion system on the Boeing 777 aircraft. Mr. Krodel is a designated engineer representative for software and also worked several years in the software quality assurance domain and is a certified lead TickIT auditor for the International Standards Organizations 9001 family of standards. He has conducted several studies relating to commercial off-the-shelf software in the avionics domain and is currently assisting in the integration of several disciplines in engine control systems. Mr. Krodel was chair of RTCA Inc.'s Special Committee (SC)190, which completed additional clarification guidance regarding DO-178B, "Software Considerations in Airborne Systems and Equipment Certification" (also known as ED-12B). This committee work included harmonization of the document with EUROCAE Working Group 52. Currently, he chairs SC205, which is taking into consideration modifications to DO-178B. He holds a master's degree in computer science from the University of Connecticut and has been a research affiliate lecturer on systems and software development in aerospace systems at the Massachusetts Institute of Technology.

RAYMOND R. LaFREY retired in 2003 as manager of the MIT Lincoln Laboratory Air Traffic Control (ATC) Mission Area. He joined Lincoln Laboratory in 1969 and began developing air traffic control technology in 1974. During 1977-1982, he led the development of the Traffic Alert and Collision Avoidance System (TCAS II) flight hardware, flight-test activities, and air-to-air coordination logic. He later led the development of a civil GPS navigation set, the Parallel Runway Monitor, and ADS-B experimental avionics and field trials. As manager of the ATC Mission Area, he oversaw the development of airport surface technology, open architecture surveillance systems, and integrated terminal and regional weather systems. He served on a Defense Science Board for Aviation Safety, the FAA Research Engineering and Development Advisory Committee, and is a member of the FAA National Airspace System Operations subcommittee. He has chaired several major FAA studies, served on two NRC studies, received awards from the FAA and a Collier Medal for his work on ADS-B. He received a B.S.E.E. (High Honors) and an M.S.E.E. at Michigan State University.

EDMOND L. SOLIDAY is an Indiana state representative serving on the transportation, commerce, energy, and technology committees. He serves on the MIT Global Airline Industry Program Advisory Group. Previously, he was employed by United Airlines for more than 35 years as a pilot, human factors instructor, flight manager, and staff executive, serving the last 11 as vice president of safety, quality assurance, and security. He has served on numerous aviation safety-related advisory boards and commissions, and he has chaired the Commercial Aviation Safety Team, the Air Transport Association Safety Council, the Star Alliance Safety Committee, and the ATA Environmental Committee. Captain Soliday formerly served on the Executive Board of the Flight Safety Foundation. Among his awards are the Bendix Trophy, the Vanguard Trophy, and the Laura Tabor Barbour International Air Safety Award. Captain Soliday previously served on NRC's Organizing Committee for the Workshop on Assessing the Research and Development Plan for the Next Generation Air Transportation System and the Steering Committee on Decadal Survey of Civil Aeronautics. He is a member of the Aeronautics and Space Engineering Board.

CORNELIA TOWNSEND is the director of Aviation Safety at Boeing Commercial Airplanes. She is responsible for leading all commercial airplane product-safety-related activities, including accident investigations, continued airworthiness efforts, and safety assessments for new and derivative airplanes. Ms. Townsend also leads Boeing's participation in industry global safety programs and commercial airplanes human factors core team, which is functionally responsible for human performance analysis and reducing human error in aviation operations. Previously, she was the chief project engineer responsible for ensuring the product integrity and safety of the 747 while improving the 747's value from the customer perspective. Her career with Boeing has included assignments in aerodynamics engineering and customer engineering. Ms. Townsend has held positions as manager, 777 program management; senior manager, 747/767/777 airplane performance, safety, certification, and test and validation; 747 airplane level functional integration team leader; and 747 deputy chief project engineer. She received a B.S. in aerospace engineering from the University of Colorado and a M.B.A. from Seattle University.

ERIC J. TUEGEL is a senior engineer at the Air Force Research Laboratory. He has been involved in structural mechanics research, especially in the area of fatigue and fracture, for more than 25 years, the last 8 years at the Air Force Research Laboratory. He spent 9 years at McDonnell Douglas Corporation (MDC). improving aircraft structural integrity analysis methods through the application of new concepts in fatigue and fracture. Prior to joining MDC, he worked on models for elasto-plastic material behavior and crack nucleation while an assistant professor of mechanical engineering at Washington University in St. Louis. Currently, Dr. Tuegel is the team leader for Structural Life Forecasting Methods in the Structural Sciences Center of the Air Vehicles Directorate. Dr. Tuegel has been a member of the American Society for Testing and Materials (ASTM) Committee E-08 on Fatigue and Fracture and the ASME since 1983. He chairs the ASTM Task Group on Variability, Statistic and Probabilistic Modeling in Fatigue for the committee. He received his B.S. in physics from Butler University and his M.S. and Ph.D. in mechanical engineering from the University of Illinois.

RAYMOND VALEIKA is an independent consultant. He advises major companies on aviation matters. He is an internationally recognized senior airline operations executive with more than 40 years of managing the maintenance operations of large airlines. Mr. Valeika recently retired from Delta Airlines as senior vice president of technical operations (TechOps), where he directed a worldwide maintenance and engineering staff of more than 10,000 professionals, maintaining a fleet of nearly 600 aircraft. Through his leadership and focus on continuous improvement of the human processes in aviation maintenance, Delta TechOps consistently rated at the top of the industry for performance benchmarks in the areas of safety, quality, productivity, and reliability. Mr. Valeika was honored with the Air Transport Association's (ATA's) Nuts & Bolts award, recognizing his leadership in the aviation industry. He has been recognized with a Humanitarian Award from the Community Mayors of New York, New Jersey, and Connecticut, and a Laurel from *Aviation Week and Space Technology* for his role in human factors training at Continental. In October 1999, Mr. Valeika received the Marvin Whitlock Award from the Society of Automotive Engineers. Most recently, the Aviation Week Group honored him with a lifetime achievement award. Previously, he held senior executive positions with Pan Am and Continental Airlines as well as Delta. He graduated from St. Louis University with a degree in aeronautical engineering. He has previously served on the Committee for the Evaluation of NASA's Fundamental Aeronautics Research Program, the Committee on Analysis of Air Force Engine Efficiency Improvement Options for Large Non-Fighter Aircraft and the Aeronautics and Space Engineering Board.

WILLIAM WHITTON is vice president of the Gulfstream Organizational Designation Authorization (ODA) Office of Gulfstream Aerospace Corporation, a subsidiary of General Dynamics. He leads the business-jet manufacturer's ODA activities, which include oversight of type certification, production certification and major repair and alterations airworthiness on behalf of the FAA. Prior to this, he served as director of aircraft system engineering where he was involved in the design, development, certification, and support of the G350, G450, G500, and G550 business jets. Prior to Gulfstream, he worked for Boeing Commercial Airplanes for 15 years, last serving as aircraft systems engineering leader on Boeing 747 programs. He has a B.S. in mathematics from the University of Puget Sound and a B.S.E.E. from Portland State University.

C

Presentation List from the 2009 Aviation Safety Program Technical Conference

NASA's Aviation Safety Program holds an annual technical conference that enables NASA researchers to meet with each other as well as engage with outsiders from other federal agencies, the public, and the aviation industry. The research presentations for the NASA Aviation Safety Program Technical Conference held on November 17-19, 2009, in McLean, Virginia, are listed below.

AIRCRAFT AGING AND DURABILITY

High Temperature Sensors and Durability of Superalloy Engine Disks

Durability of Engine Hot Section
Development of Ductile Coatings to Protect Advanced Powder Metallurgy Alloy Turbine Disks from Hot Corrosion Attack
Effect of Dwells on Notch Fatigue Life in Nickel-based Disk Alloys
Application of Computational Microstructure Modeling to 3rd Generation Nickel-Based Turbine Disk Alloys

Metallic and Integral Structures

Developing a Practical Nonlinearity Parameter Measurement Technology to Assess Fatigue Damage in Structures
Digital Image Correlation for Nano-Scale Fracture
Measuring Residual Stress and Developing Reliable Fatigue Life Assessments

Bonded Structures

Non-Linear UT Assessment of Adhesive Bonds
Laser Surface Preparation and Bonding of Aerospace Structural Composites
Preliminary Results of Aging

Wiring

Investigation of Dielectric and Thermal Properties of Wire Insulation Polymers for Development of Capacitive Nondestructive Evaluation
Dynamic Wiring Fault Detection Research
An Instrumented Crimp Tool for Electrical Termination Inspection

INTEGRATED INTELLIGENT FLIGHT DECK

Flight Deck System Research—Multi-Disciplinary Solution Concepts

RDT&E of a Robust Automation-Human System Concept of Operations
RDT&E of a "Better-than-Visual" Display and Decision Support Concept of Operations

Discipline-Specific Research

Integrated Alerting & Notification: ALARMS—Alerting and Reasoning Management System
Testing and Validation of a Psychophysically Defined Metric of Display Clutter

Foundational Research

Forward-Looking Sensing of Aviation Hazards
Characterizing and Classifying the Functional State of Operators
Context-Sensitive Multi-Modal Interfaces

Design Tools

Advanced Computational Methods for the Design of Automated Systems
Next Generation Automation Interaction Design Tools
Design Advisor Tool
Surface Map Traffic Intent Displays and Net-Centric Data-Link Communications for NextGen

INTEGRATED RESILIENT AIRCRAFT

Resilient Flight Control

Advances in Adaptive Control Methods for Resilient Flight Control
An Adaptive Control Technology for Flight Safety in the Presence of Actuator Anomalies and Damage
Composite Model Reference Adaptive Control
Analytical Validation Tools for Safety Critical Aerospace Systems
Internal Algorithm Monitor for Adaptive Systems
Constraining the Flight Maneuvering Envelope for Structurally Damaged Aircraft
Damage-Resilient Flight Planning and Guidance System for Safe, Collaborative Emergency Management
Modeling the Human Pilot Controlling Systems with Rapidly Changing Dynamics
Flight Test Evaluation of the Smart-Cue and Smart-Gain Concepts to Alleviate Pilot-Vehicle System Loss of Control

Technology Transition

AirSTAR L1 Adaptive Control
Flight Test Evaluation of Adaptive Control with AirSTAR: Techniques and Capabilities
Technology Challenges and Metrics for Certification
Adaptive Control Validation in Piloted Simulator
Adaptive Control Validation in F-18

INTEGRATED VEHICLE HEALTH MANAGEMENT

Detection

Microwave Blade Tip Clearance Sensor for Propulsion Health Monitoring
High Temperature Wireless Sensor Systems
Multifunctional Sensing using Fiber Bragg Gratings
Integrated Large Area Sensor Actuator Network Technology for Structural Health Monitoring
Data Mining for Fleet-wide Health Monitoring
Event Cube: An Organized Approach for Mining and Understanding Anomalous Aviation Events

Diagnosis

On-Board Model-Based Aircraft Engine Performance Estimation for IVHM Applications
Lightning Damage Diagnosis Research for Composite Aircraft
Probabilistic Methods for Diagnosis of Aircraft Systems
Diagnostics of Avionics Systems using Causal Methods
New Algorithms for Diagnosis of Multiple Faults
Vehicle Level Reasoning for Integrated Vehicle Health Management

VERIFICATION AND VALIDATION

V&V of Flight-Critical Systems
Argument-based Safety Assurance
Formal Verification using Hybrid Abstraction
Guaranteeing Controls Properties from Specifications to Code
Hierarchical Component-based Framework for formal V&V of complex aerospace software
Software Intensive Systems

D

Safety R&T Challenges from the *Decadal Survey of Civil Aeronautics*

The panels of the *Decadal Survey of Civil Aeronautics* ranked the national priority of research and technology (R&T) safety challenges on a scale of 1 to 9, with 9 being the highest score.[1] The 39 R&T challenges listed in Table D.1 were rated "9" for safety priority. In addition, the panels ranked each of the challenges with respect to their suitability for research by NASA. None of the safety challenges received a "9," but many were rated "6" or higher, as shown in the table.

[1] National Research Council, *Decadal Survey of Civil Aeronautics: Foundation for the Future,* The National Academies Press, Washington, D.C., 2006.

TABLE D.1 The *Decadal Survey's* Top-Ranked R&T Challenges for Safety Importance (those challenges scoring a "9" for safety) (9 is highest score; 1 is lowest)

R&T Challenge by Discipline/Area	Suitability for NASA to Address in its R&T Programs
Propulsion and Power	
Intelligent engines and mechanical power systems capable of self-diagnosis and reconfiguration between shop visits	6
Improved propulsion system tolerance to weather, inlet distortion, wake ingestion, bird strike, and foreign object damage	3
Materials and Structures	
Innovative load suppression and vibration and aeromechanical stability control	7.5
Innovative high-temperature metals and environmental coatings	7.5
Integrated vehicle health management	7
Novel coatings	6
Next-generation nondestructive evaluation	4
Aircraft hardening	4
Intelligent and Autonomous Systems, Operations and Decision-making, Human Integrated Systems, and Networking and Communications	
Methodologies, tools, and simulation and modeling capabilities to design and evaluate complex interactive systems	6
New concepts and methods of separating, spacing, and sequencing aircraft	6
Appropriate roles of humans and automated systems for separation assurance, including feasibility and merits of highly automated separation assurance systems	6
Affordable new sensors, system technologies, and procedures to improve the prediction and measurement of wake turbulence	6
Interfaces that assure effective information sharing and coordination among ground-based and airborne human and machine agents	6
Vulnerability analysis as an integral element in the architecture design and simulations of the air transportation system	6
Transparent and collaborative decision-support systems	6
Using operational and maintenance data to assess leading indicators of safety	6
Interfaces and procedures that support human operators ineffective task and attention management	6
Autonomous flight monitoring and manned and unmanned aircraft	6
Technologies to enable refuse-to-crash and emergency auto-land systems	6
Feasibility of deploying an affordable broad-area, precision-navigation capability compatible with international standards	4.5
Change management techniques applicable to the US air transportation system	2.5
Provably correct protocols for fault-tolerant aviation communications systems	2
Comprehensive models and standards for designing and certifying aviation networking and communications systems	2
Dynamics, Navigation, Control, and Avionics	
Advanced guidance systems	7.5
Aerodynamics and vehicle dynamics via closed loop flow control	7.5
Fault-tolerant and integrated vehicle health management systems	7.5
Distributed decision-making, decision making under uncertainty, and flight-path planning and prediction	6

continues

TABLE D.1 Continued

R&T Challenge by Discipline/Area	Suitability for NASA to Address in its R&T Programs
Intelligent and adaptive flight control techniques	6
Improved on-board weather systems and tools	6
Human-machine integration	6
Synthetic visions and enhanced vision systems	6
Advanced communications, navigation, and surveillance technology	4.5
Secure network-centric avionics architectures and systems to provide low cost, efficient, fault-tolerant, on-board communications systems for data link and data transfer	4.5
Safe operation of unmanned air vehicles in the national airspace	4.5
More efficient certification processes for complex systems	2
Design, development, and upgrade processes for complex, software-intensive systems, including tools for design, development, and verification and validation	1.5
Aerodynamics and Aerocoustics	
Aerodynamics robust to atmospheric disturbances and adverse weather conditions, including icing	4.5
Accuracy of wake vortex prediction, and vortex detection and mitigation techniques	4.5

E

NASA-Identified R&T Challenges from the *Decadal Survey of Civil Aeronautics*

The *Decadal Survey of Civil Aeronautics*[1] identifies numerous research and technology (R&T) challenges within the field of Aeronautics. In briefings to the committee, NASA identified the R&T challenges relevant to the work being pursued in its various aviation safety-related research activities. These challenges are listed in the various tables below. The "NASA Relevance and Suitability" scores are "Why NASA Composite Score" in the *Decadal Survey*, and they reflect a combination of four distinct factors:

- Supporting infrastructure
- Mission alignment
- Lack of alternative sponsors
- Appropriate level of risk

Due to the subjective nature of the scoring and differences between panels, the *Decadal Survey* cautions readers against direct comparison of the scores for challenges from different panels (each panel was assigned a letter, A-E).

[1] National Research Council, *Decadal Survey of Civil Aeronautics: Foundation for the Future,* The National Academies Press, Washington, D.C., 2006.

TABLE E.1 The *Decadal Survey's* Top-Ranked Safety R&T Challenges Applicable to New Operations (9 is highest score; 1 is lowest)

R&T Challenge by Discipline/Area	NASA Relevance and Suitability
Dynamics, Navigation, Control, and Avionics	
(D1) Advanced guidance systems	7.5
(D2) Distributed decision-making, decision making under uncertainty, and flight-path planning and prediction	6
(D6) Improved on-board weather systems and tools	6
(D8) Human-machine integration	6
(D9) Synthetic vision and enhanced vision systems	6
(D7) Advanced communications, navigation, and surveillance technology	4.5
Intelligent and Autonomous Systems, Operations and Decision-making, Human Integrated Systems, and Networking and Communications	
(E1) Methodologies, tools, and simulation and modeling capabilities to design and evaluate complex interactive systems	6
(E2) New concepts and methods of separating, spacing, and sequencing aircraft	6
(E3) Appropriate roles of humans and automated systems for separation assurance, including feasibility and merits of highly automated separation assurance systems	6
(E4) Affordable new sensors, system technologies, and procedures to improve the prediction and measurement of wake turbulence	6
(E5) Interfaces that assure effective information sharing and coordination among ground-based and airborne human and machine agents	6
(E8a) Transparent and collaborative decision-support systems	6
(E8c) Interfaces and procedures that support human operators ineffective task and attention management	6

SOURCE: Selected from the "Prioritization of R&T Challenges" tables in the *Decadal Survey.*

TABLE E.2 The *Decadal Survey's* Top-Ranked Safety R&T Challenges Applicable to Hazardous Conditions and the IIFD and IRAC Technical Plans (9 is highest score; 1 is lowest)

R&T Challenge by Discipline/Area	NASA Relevance and Suitability
Intelligent and Autonomous Systems, Operations and Decision-making, Human Integrated Systems, and Networking and Communications	
(E4) Affordable new sensors, system technologies, and procedures to improve the prediction and measurement of wake turbulence	6
Dynamics, Navigation, Control, and Avionics	
(D6) Improved on-board weather systems and tools	6
(D7) Advanced communications, navigation, and surveillance technology	4.5
Aerodynamics and Aerocoustics	
(A6) Aerodynamics robust to atmospheric disturbances and adverse weather conditions, including icing	4.5

SOURCE: Selected from the "Prioritization of R&T Challenges" tables in the *Decadal Survey.*

TABLE E.3 The *Decadal Survey's* Top-Ranked Safety R&T Challenges Applicable to Loss and Control (9 is highest score; 1 is lowest)

R&T Challenge by Discipline/Area	NASA Relevance and Suitability
Dynamics, Navigation, Control, and Avionics	
(D1) Advanced guidance systems	7.5
(D2) Distributed decision-making, decision making under uncertainty, and flight-path planning and prediction	6
(D3) Intelligent and adaptive flight control techniques	6
(D13) More efficient certification processes for complex systems	2
(D14) Design, development, and upgrade processes for complex, software-intensive systems, including tools for design, development, and verification and validation	1.5

SOURCE: Selected from the "Prioritization of R&T Challenges" tables in the *Decadal Survey*.

TABLE E.4 The *Decadal Survey's* Top-Ranked Safety R&T Challenges Applicable to Durable Aircraft Structures and Systems According to the AAD Technical Plan (9 is highest score; 1 is lowest)

R&T Challenge by Discipline/Area	NASA Relevance and Suitability
Materials and Structures	
(C1) Integrated vehicle health management	7
(C6a) Innovative high-temperature metals and environmental coatings	7.5
(C14) Next-generation nondestructive evaluation	4

SOURCE: Selected from the "Prioritization of R&T Challenges" tables in *Decadal Survey*.

TABLE E.5 The *Decadal Survey's* Top-Ranked Safety R&T Challenges Applicable to On-Board System Failures and Faults According to the IVHM Technical Plan (9 is highest score; 1 is lowest)

R&T Challenge by Discipline/Area	NASA Relevance and Suitability
Propulsion and Power	
(B3) Intelligent engines and mechanical power systems capable of self-diagnosis and reconfiguration between shop visits	6
Materials and Structures	
(C1) Integrated vehicle health management	7
Intelligent and Autonomous Systems, Operations and Decision-making, Human Integrated Systems, and Networking and Communications	
(E1) Methodologies, tools, and simulation and modeling capabilities to design and evaluate complex interactive systems	6
(E8b) Using operational and maintenance data to assess leading indicators of safety	6
Dynamics, Navigation, Control, and Avionics	
(D5) Fault-tolerant and integrated vehicle health management systems	7.5

SOURCE: Selected from the "Prioritization of R&T Challenges" tables in the *Decadal Survey*.

TABLE E.6 The *Decadal Survey's* Top-Ranked Safety R&T Challenges Applicable to Analyzing Complex Systems for Safety (9 is highest score; 1 is lowest)

R&T Challenge by Discipline/Area	NASA Relevance and Suitability
Intelligent and Autonomous Systems, Operations and Decision-making, Human Integrated Systems, and Networking and Communications	
(E1) Methodologies, tools, and simulation and modeling capabilities to design and evaluate complex interactive systems	6
(E2) New concepts and methods of separating, spacing, and sequencing aircraft	6
(E3) Appropriate roles of humans and automated systems for separation assurance, including feasibility and merits of highly automated separation assurance systems	6
(E5) Interfaces that ensure effective information sharing and coordination among ground-based and airborne human and machine agents	6
(E6) Vulnerability analysis as an integral element in the architecture design and simulations of the air transportation system	6
(E8a) Transparent and collaborative decision-support systems	6
(E8c) Interfaces and procedures that support human operators ineffective task and attention management	6
Dynamics, Navigation, Control, and Avionics	
(D1) Advanced guidance systems	7.5
(D2) Distributed decision-making, decision making under uncertainty, and flight-path planning and prediction	6
(D6) Improved on-board weather systems and tools	6
(D8) Human-machine integration	6
(D9) Synthetic visions and enhanced vision systems	6
(D7) Advanced communications, navigation, and surveillance technology	4.5

SOURCE: Selected from the "Prioritization of R&T Challenges" tables in the *Decadal Survey.*

TABLE E.7 The *Decadal Survey's* Top-Ranked Safety R&T Challenges Applicable to Safety-Related Research in the Fundamental Aeronautics Program (9 is highest score; 1 is lowest)

R&T Challenge by Discipline/Area	NASA Relevance and Suitability
Aerodynamics and Aerocoustics	
(A4a) Aerodynamic designs and flow control schemes to reduce aircraft and rotor noise	6
(A6) Aerodynamics robust to atmospheric disturbances and adverse weather conditions, including icing	4.5
Materials and Structures	
(C8) Structural innovations for high-speed rotorcraft	7.5

SOURCE: Selected from the "Prioritization of R&T Challenges" tables in the *Decadal Survey.*

F

Acronyms

AAD	Aircraft Aging and Durability
AFRL	Air Force Research Laboratory
AIRA	Aircraft Icing Research Alliance
ARMD	Aeronautics Research Mission Directorate
ARRA	American Recovery and Reinvestment Act
ASEB	Aeronautics and Space Engineering Board
ASIAS	Aviation Safety Information Analysis and Sharing
ATM	Air Traffic Management
CAST	Commercial Aviation Safety Team
CICTT	CAST/ICAO Common Taxonomy Team
DOD	Department of Defense
DOT	Department of Transportation
FAA	Federal Aviation Administration
FAST	Facilitated Access to the Space Environment for Technology
FDR WG	Flight Deck Research Working Group
FEM	Finite element method
HIWC	High ice water content
IATA	International Air Transport Association
ICAO	International Civil Aviation Organization
IIFD	Integrated Intelligent Flight Deck
IRAC	Integrated Resilient Aircraft Control
IVHM	Integrated Vehicle Health Management
IWP	Integrated Work Plan

JIMDAT	Joint Implementation Measurement Data Analysis Team
JPDO	Joint Planning and Development Office
LEWICE	Lewis ICE accretion program software
NACA	National Advisory Committee for Aeronautics
NASA	National Aeronautics and Space Administration
NDE	Nondestructive evaluation
NDI	Nondestructive inspection
NextGen	Next Generation Air Transportation System
NOAA	National Oceanic and Atmospheric Administration
NRA	NASA Research Announcement
NRC	National Research Council
NTSB	National Transportation Safety Board
R&D	Research and development
R&T	Research and technology
RTT	Research Transition Team
SBIR	Small Business Innovation Research
SRM	Safety risk management
TRB	Transportation Research Board
UAS	Unmanned Aircraft Systems
USAF	United States Air Force
V&V	verification and validation